ENERGY CRISES

IN PERSPECTIVE

ENERGY CRISES

IN PERSPECTIVE

JOHN C. FISHER

A WILEY-INTERSCIENCE PUBLICATION

JOHN WILEY & SONS, New York • London • Sydney • Toronto

Library of Congress Cataloging in Publication Data:

Fisher, John Crocker.
  Energy crises in perspective.

  "A Wiley-Interscience publication."
  1. Power Resources.   I. Title.

HD9540.6.F57   333.7   73-17049
ISBN 0-471-26141-6

2 3 4 5 6 7 8 9 10

PREFACE

This book was written for people with a general or professional interest in world energy activities who might like to see an undistorted picture of the world energy situation as it is, how it got there, and where it is going. I had in mind investors, security analysts, environmentalists, energy industry managers, economists, engineers, writers, lawyers, university professors, students, government agency employees, and others with an interest in being able to recognize and understand significant events in the world energy drama, and perhaps to anticipate some of them. The treatment is quantitative in several places and slightly technical in others, but no special qualifications beyond literacy and attention are required of the reader.

World energy activities, from discovery through recovery, transportation, refining, distribution, and use, are molded by tremendous outside forces: forces of supply and demand, themselves shaped by interrelated political, social, economic, and technological pressures. My approach has been to strip away the camouflage from major world energy activities and from major factors affecting them, to examine their structures and interactions, and to see how they fit together and behave as a whole. The goal has been to develop a conceptual framework for thought and action, in which today's energy crisis, as others before and yet to come, can be appreciated in perspective.

The writing of the book was sponsored by the General Electric Company, to whom I am grateful for providing the opportunity, with the instruction that I write as objectively as I could. I have done my best in this respect. In no way does the conceptual framework that is developed represent a "General Electric view" of the situation. My approach has been analytical and pedagogical,

v

whereas corporations tend to be commercial and competitive. Each approach is proper in its place, but they can evoke different views, as for example my projections compared with various industry targets for future electrification.

A number of people have made valuable suggestions and criticisms of the work as it progressed, and I would like to acknowledge and thank M. H. Bensky, K. P. Cohen, C. Concordia, N. J. Crum, C. W. Elston, H. B. Finger, A. M. Gurewitsch, H. R. Hill, E. B. Hutchins, G. G. Leeth, J. A. Long, M. J. McNelly, P. M. Murphy, D. R. Plumley, J. B. Tice, D. M. Willyoung, and J. F. Young for their help in this respect. In addition, I give special thanks to E. J. Schmidt and T. O. Paine who focused my attention on the energy situation and who encouraged and supported my efforts to understand it.

JOHN C. FISHER

September 1973
Wilton, Connecticut

CONTENTS

ENERGY CRISES

IN PERSPECTIVE

CHAPTER ONE

THE ROLE OF ENERGY

Energy can provide for a wide range of personal needs and wants. In the form of food, it is essential to life. Beyond food, some of the most significant personal uses in terms of energy requirements include cooking, comfort heat, illumination, personal transportation, hot water, refrigeration, entertainment, and comfort cooling. These uses extend far beyond supplying the bare essentials for life, and they provide increasingly for comfort and convenience. Industry and agriculture utilize energy for process heat, for electrochemistry, and for work in fields, factories, and mines.

The pattern of energy consumption in industrialized societies differs substantially from that in nonindustrialized societies. Nonindustrialized societies still are heavily dependent on the traditional energy sources of antiquity—local solar energy that is made available through the agencies of food, work animal feed, fuel wood, fuel dung, agricultural wastes, windpower, and direct waterpower. Field work is largely accomplished by the power of human and animal muscles, and its energy sources are from food and animal feed. Per capita consumption of energy is very small, only a few times the food energy required to sustain life. In contrast, industrialized societies consume large quantities of fossil fuel and electricity, the fuel consisting of coal, oil, and natural gas, and the electricity generated partly from fuel and partly from falling water. Fossil fuels, and to a lesser extent electricity, are shipped long distances from their points of origin to their points of consumption. Per capita consumption of energy is as much as a hundred times that contained in food.

3

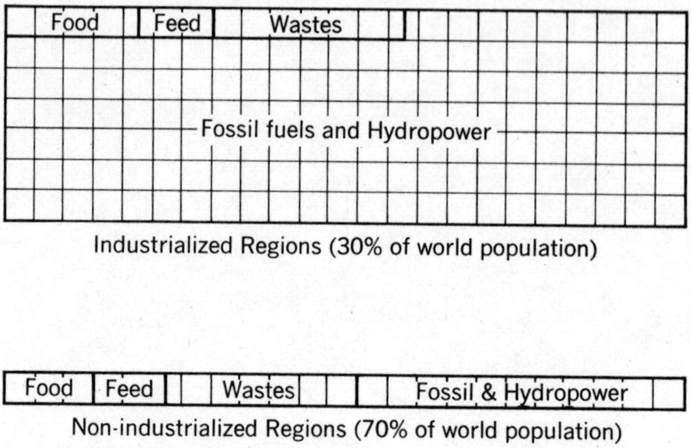

Figure 1.1. Per capita consumption of energy in 1970 (1.1). Each square represents 1 million Btu. The traditional energy sources of antiquity are food, work animal feed, and nonmineral fuels such as agricultural wastes, dung, and wood. Fossil fuels and hydropower provide a twelve-fold increase in energy for the industrialized regions, compared with a two-fold increase for the nonindustrailized regions.

Figure 1.1 illustrates the tremendous per capita consumption of fossil fuels and hydropower by the industrialized regions of the world—Anglo-America, Western Europe, the Soviet Bloc and Oceana—compared with the rest of the world. The nonindustrialized regions, with 70 percent of world population, consume only enough fossil fuels and hydropower to double the traditional supply of antiquity. The industrialized regions, with 30 percent of world population, consume so much as to dwarf the traditional supplies.

Technology is a major factor in the effectiveness and economics of energy utilization. Its influence on agriculture and food production is of particular concern to nonindustrialized societies where food and feed are primary energy sources. Its influence on the recovery, processing, and utilization of inanimate fuel and energy is of particular concern to industrialized societies. This book is directed primarily to problems and opportunities in the second of these areas, the utilization of inanimate energy, with particular emphasis on electrical energy.

A comprehensive analysis of the patterns of energy consumption in the United States has been made by the Stanford Research Institute for the U.S. Office of Science and Technology (1.2). This study is the first of its kind to attempt so detailed an analysis, and for want of adequate data it sometimes was necessary for the authors to make estimates, yet overall I believe it to be the best study available. It covers the years 1960–1968 and investigates the transportation, residential, commercial, and industrial energy-use sectors to varying depths. Key data are summarized in Appendix 1. The sectors themselves consumed energy in the following proportions in 1968:

Industrial	39.9%
Transportation	25.6
Residential	20.6
Commercial	13.9
	100.0

The transportation sector includes farm work, other nonfactory work, and vehicle air conditioning in addition to simple movement of people and goods, but these end uses were not separately determined. The other sectors were treated in more detail, and a number of significant end uses were identified and separately quantified. Table 1.1 lists these uses in order of the amount of energy consumed. These figures show clearly that most energy consumption is associated with the various activities of the working population in industry and commerce, including the associated transportation of goods and of workers to

TABLE 1.1 SIGNIFICANT END USES FOR ENERGY (U.S.A., 1968)

End Use	Percent
Transportation	25.6
Space heating	20.8
Industrial process steam	14.6
Industrial direct heat	11.0
Industrial drive	10.3
Water heating	4.2
Refrigeration	3.0
Lighting	2.8
Air conditioning	1.5
Electrolytic processes	1.5
Cooking	1.4
Television	0.7
Clothes drying	0.4
Other (electric)	2.2
	100.0

and from work. A much smaller portion is associated with nonworking activities, including pleasure transportation and various residential energy uses.

Between 1960 and 1968, the consumption of energy in the United States increased at the average rate of 4.3 percent annually, with some uses of energy growing more rapidly than others. The major uses can be divided into two classes: relatively slowly growing basic uses that have been fully assimilated by the American economy and society, and new uses that are growing more rapidly in the process of becoming assimilated.

Basic uses of energy include transportation, space heating, process steam, direct heat, industrial drive, water heating, electrolytic processes, and cooking. In aggregate, they grew at an average rate of 3.9 percent. These uses have been part of the American way of life for centuries, except for electrolytic processes which have been so for decades. Every working adult requires transportation, which has come to mean a personal automobile. Every new house or apartment has its share of space heating, water heating, and cooking; every new commercial establishment has the same; and every new factory has its share of space heating, process steam, direct heat, industrial drive, and electrolytic processing. The basic energy uses tend to grow in proportion to the everyday needs of housekeeping, officekeeping, and factorykeeping.

The progressive movement of working people from farms to factories and offices, and more recently of women from homes to factories and offices, has of necessity been accompanied by a growth of industrial and commercial establishments and transportation facilities and in their basic energy requirements, at a rate exceeding the growth in total population. The extra commercial space had to be heated and plumbed with hot water. The extra industrial space had to have appropriate quantities of process steam, direct heat, industrial drive, and electrolytic processes. The new workers had to have transportation to and from work. In quantitative terms, the basic energy uses associated with the commercial, industrial, and transportation sectors would be expected to increase in proportion to the number of nonfarm workers, even if there were no increase in the average energy content of the goods and services produced. Any increase in average energy content, such as might be expected for manufacturing a modern automobile in comparison with its lighter, simpler ancestor, would lead to additional energy consumption.

This concept can be put to use. In the years 1960–1968, the nonfarm working population (specifically, the population employed in nonagricultural establishments including manufacturing, wholesale and retail trade, government, services, transportation, public utilities, finance, insurance, real estate, contract construction, and mining) grew more than twice as fast as the general population, increasing at the rate of 2.9 percent annually compared to the general population which grew at 1.3 percent annually (1.3). The basic energy requirements of commerce and industry would be expected to grow at this same rate, provided the energy content per unit output of goods and services remained relatively unchanged. In actuality, the commercial, industrial and transportation sectors increased their basic energy consumption at an average rate of 3.8 percent, corresponding to a per-working-capita growth in basic energy usage of 0.9 percent per year. This extra growth indicates a progressively higher energy content of the goods and services produced. Summarized, these annual growth rates are

Nonfarm working population	2.9 percent
Energy content per unit of goods and services	0.9 percent
	———
Basic energy consumed by transportation, commercial, and industrial sectors	3.8 percent.

The basic energy consumption of the residential sector, including space heating, water heating, and cooking, would be expected to grow in proportion to the total population, plus any per capita increase due to increasing affluence and changing life styles. The data show that in the years 1960–1968, the residential basic energy consumption increased on average 4.1 percent per year, compared to the population growth of only 1.3 percent, showing a substantial affluence-related growth in energy consumption of 2.8 percent annually. This increase is mostly attributable to space heating and water heating, and practically none to cooking. Summarized, these annual growth rates are

Population	1.3 percent
Per capita residential basic energy	2.8 percent
	———
Basic energy consumed by residential sector	4.1 percent.

The new uses of energy include refrigeration, air conditioning, television, clothes drying, and a number of electric and electronic uses including lighting (1.4). These uses are in the process of finding their way into society. Many of them depend on electricity, without which they would be prohibitively expensive or technically impossible. Experience suggests that most will, in time, become universally accepted and join the other basic uses for energy. In aggregate, these new uses accounted for about 11 percent of energy consumption in the United States in 1968 and were growing at about 8.5 percent per year as they moved on toward full assimilation. This annual growth rate is

New energy consumed by all sectors	8.5 percent.

Five key annual growth rates have been identified in the preceding paragraphs: (1) nonfarm working population, (2) energy content per unit of goods and services, (3) population, (4) per capita residential basic energy, and (5) new energy consumed by all sectors. By estimating the future evolution of these growth rates it is possible to project the future consumption of energy in the United States. But first let us examine and review the various sources from which the energy might come.

CHAPTER TWO

THE SOURCES OF ENERGY

A number of different sources have provided significant energy inputs to the United States at one time or another. In approximate order of their historic development, they can be classified under the headings of solar energy, fossil fuels, and nuclear fuels:

SOLAR ENERGY: conversion via
Fuel wood
Work animal feed
Direct windpower
Direct waterpower
Hydroelectricity

FOSSIL FUELS: combustion of
Coal
Petroleum
Natural gas

NUCLEAR FUELS: fission of
Uranium
Thorium

Solar energy is dilute, but large in magnitude and unlimited in time. Fossil fuels are concentrated and cheap, but are exhaustible and can become exhausted after several centuries. Nuclear fuels are practically inexhaustible,

TABLE 2.1 SOURCES OF ENERGY FOR
THE UNITED STATES IN 1970

Source	Percent
Fossil fuel	
Coal	20
Petroleum	41
Natural gas	33
Solar energy	
Hydroelectricity	4
Other	
Miscellaneous	2
	100

particularly if breeder reactors are able to utilize the common isotopes of uranium and thorium.

Broadly speaking, for the industrialized societies the years of significance for fuel wood, work animal feed, direct windpower, and direct waterpower have passed; and the years of significance for nuclear fuels are just beginning. The energy sources of current significance are hydroelectricity and the fossil fuels. In the United States in 1970, 94 percent of the energy consumed came from fossil fuels, 4 percent from the fuel equivalent of hydroelectric power, and the remaining 2 percent from various minor sources, including a rapidly growing one-third of 1 percent from nuclear fuels. These figures are summarized in Table 2.1.

Other sources of potential significance for large-scale energy production include:

SOLAR ENERGY: conversion via
   Fuel crops

FOSSIL FUEL: combustion of hydrocarbons from
   Oil shale
   Tar sand

Sources of energy are judged to be potentially significant where the available quantities are large and where technological and economic considerations show that the costs are competitive or close to competitive. Other potential sources such as tidal power, geothermal power, fusion power, and trash combustion are judged not to be significant for large-scale energy production because of limited availability or because of economic or technological barriers, although they may have limited applications at special locations or in special situations.

The 1970 emphasis on fossil fuels in the United States represents a strong shift from the 1850 emphasis on wood (for heat) and animal feed (for farm work and transportation), as summarized in Table 2.2. The years from 1850 to 1970 saw several major substitutions of new energy forms for the old, as shown graphically in Figure 2.1. Fuel wood, used primarily for heating, was largely replaced by coal between 1850 and 1910. Since 1910, coal has been progressively replaced by fluid hydrocarbons. Work animal feed, used primarily for motive power in transportation and on farms, was partially replaced by coal in the late 1800's and early 1900's. Then, as the country adopted automobiles and tractors and as railroads converted to oil, both animal feed and railroad coal were largely replaced by distillate motor fuels in the years 1920–1950. Direct windpower and waterpower were replaced by hydroelectricity in the years 1890–1940. These trends are derived from basic historic data for energy consumption in the United States between 1850 and 1970, collected and analyzed in Appendix 2.

Thus far I have reviewed the qualitative mix of energy sources as it has evolved to its present emphasis on fossil fuels. Now I would like to consider some of the quantitative aspects of energy consumption. Miners deal in tons of coal, oil men in barrels of oil, gas men in cubic feet of gas, and electric utility men in kilowatt-hours of hydroelectricity. Some uniform standard of measurement is required for comparing energies. Energy sources that customarily are used for the production of heat, either for an end use such as cooking or for an intermediary use such as generating electricity, can be quantified by the amounts of heat they are capable of generating. More specifically,

TABLE 2.2   SOURCES OF ENERGY FOR
THE UNITED STATES IN 1850

Source	Percent
Solar energy	
Fuel wood	64
Work animal feed	22
Wind and water	7
Fossil fuel	
Coal	7
	100

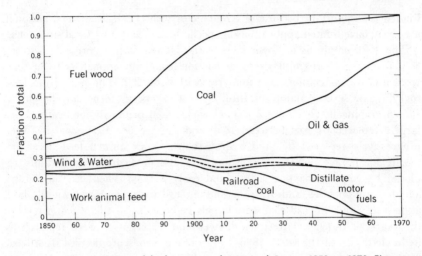

Figure 2.1. Segmentation of fuel input to the United States, 1850 to 1970. Five great energy substitutions have occurred during the past 120 years. Fuel wood, used primarily for heating, was largely replaced by coal between 1850 and 1910. Coal has since been progressively replaced by gas and oil. Work animal feed was partially replaced by railroad coal in the late 1800's and early 1900's. Then with a shorter time constant, animal feed and railroad coal both were replaced by distillate motor fuels between 1920 and 1950. Direct wind and water power were replaced by hydroelectricity in the years 1890 to 1940. A sixth substitution, not explicitly shown in the figure, is a steady increase in the proportion of fossil fuel converted to electricity prior to consumption.

the numerical values for fossil fuels, fuel wood, and animal feed are the amounts of heat they would generate during combustion. The values for nuclear fuels are the amounts of heat generated by nuclear fission in electric power plants.

Hydroelectricity presents a special problem. Its energy content can be measured either by the amount of heat it would generate in an electric heater or by the larger amount of heat that would be required to generate the same amount of electricity in a fuel-burning powerplant—this being the only alternative for obtaining electricity. I use the second of these measures, the fuel equivalent of hydroelectricity, because I believe it more accurately reflects hydroelectricity's economic significance. Similarly, direct windpower and direct waterpower present a special problem. Their energy content could be measured either by the amount of frictional heat they could generate or by the larger amount of feed energy that would be required to generate the same amount of work by work animals—this being the only alternative in the old days. Here too I use the second measure, the feed equivalent of windpower and waterpower, because I believe it more accurately reflects their economic significance.

Heat can be measured in terms of British thermal units, or Btu for short. One Btu is the amount of heat it takes to warm up 1 pound of water (approximately 1 pint of water) 1°F. Approximate energy contents for different energy sources, measured in Btu are

Source	Approximate Energy Content
Anthracite coal	12,700 Btu per pound
Bituminous coal	12,300 Btu per pound
Crude oil	5,800,000 Btu per barrel (42 gallons)
Natural gas	1,035 Btu per cubic foot
Hydroelectricity Fuel equivalent (1970)	10,500 Btu per kilowatt-hour

The United States consumes so much energy that the annual number of Btu is very large. To bring such large numbers down to size, I define a C-unit as

$$1 \ C = 10^{16} \ \text{Btu}.$$

One C of heat is approximately the heat that would be generated by burning 400 million tons of coal, the approximate amount of coal consumed in the United States each year for the past half century. (Hence the use of the letter C

TABLE 2.3   SOURCES OF ENERGY FOR THE UNITED STATES IN 1970

Source	Conventional Quantity		Energy Content	
			$C(=10^{16} \text{ Btu})$	%
Fossil fuel				
Coal	525	million tons	1.28	20
Petroleum	5.36	billion barrels	2.65	41
Natural gas	21.4	trillion cubic feet	2.13	33
Solar energy				
Hydroelectricity	253	billion kilowatt-hours	0.26	4
Other				
Miscellaneous			0.13	2
			6.45	100

TABLE 2.4   SOURCES OF ENERGY FOR
THE WORLD IN 1970

	Energy Content	
Source	$C(=10^{16}$ Btu$)$	%
Fossil fuel		
Coal	6.5	30
Petroleum	7.7	36
Natural gas	3.8	18
Solar energy		
Hydroelectricity	1.3	6
Traditional		
Wood, waste, feed	2.2	10
	21.5	100

for the unit.) It is also approximately the heat required to warm up Lake Michigan 1°F, for there are just about 10^{16} pints of water in the lake (2.1).

Now it is possible to quantify the various energy inputs as shown in Table 2.3. The percentage figures in the last column were used previously in Table 2.1 without a full explanation of their origin. Here we see how they were obtained. Overall in 1970, the energy input to the United States amounted to

TABLE 2.5   PHYSICAL QUANTITIES OF FUELS CONSUMED IN 1970

Energy Source	World Quantity		United States Quantity		United States Per Capita Tons
	Millions of Tons	Cubic Miles	Millions of Tons	Cubic Miles	
Coal (crushed)	2657	0.72	525	0.14	2.6
Petroleum	2350	0.59	800	0.20	4.0
Natural gas	820		460		2.3
One atmosphere		260		145	
65 atmospheres		4.00		2.25	
Liquified		0.43		0.24	
Falling water					
Falling 100 feet		3400		730	

6.45 C. (Subtracting out 0.26 C of hydroelectricity and adding in 0.39 C of fossil fuels used for nonenergy purposes, the overall fossil fuel input to the United States in 1970 was 6.58 C.)

World energy consumption in the same year amounted to about 21.5 C including 19.3 C of mineral fuels and hydroelectricity (2.2) and an estimated 2.2 C of traditional fuels as shown in Table 2.4.

Fossil fuels and hydroelectricity are often produced in remote regions where production costs are low, then transported over substantial distances to reach the populations that consume them. The physical volume and mass of fossil fuel that must be moved from points of production to points of consumption each year is very large, exceeding all the rest of the goods of the world combined. As an aid to appreciating the quantities of fuel that enter the world economy these days, we can look at the physical volumes on both a total and a per capita basis, as shown in Table 2.5. Substantial transportation facilities, including railroads, ships, pipelines, and high-voltage transmission lines, are required for moving these quantities of fuel and hydroelectricity. The cost of transportation is a significant factor in the overall cost of energy for in-dustrialized societies, for it is the combination of production and trans-portation costs that determines total cost, and the two components are roughly equal in magnitude.

CHAPTER THREE

PROJECTED ENERGY CONSUMPTION

In projecting energy consumption, let us first consider just the United States, which consumes much of the world's energy and is to some extent representative of the industrialized regions of the world. Before projecting into the future, let us look back at what has happened in the past.

Figure 1.1 has shown the tremendous difference in per capita energy consumption between the industrialized and the nonindustrialized regions in modern times. The preindustrialized United States seems to have been an exception, and appears never to have passed through a period of low per capita energy consumption once the European settlers arrived. The first settlers found a continent completely covered with virgin forest. This standing timber was available as fuel wood for the cutting, and the small population began consuming energy at a high per capita rate, far beyond the level of energy consumption in traditional societies, as soon as they stepped ashore. In 1850, for example, per capita energy consumption was nearly half as great as it is today. Fortunately for us, coal, oil, and gas were discovered and put to use as the forests were cut down and the population grew, and the transition from fuel wood to fossil fuels was smooth. Figure 3.1 shows the slow growth in per capita energy consumption in the United States from 1850 to 1970, starting from the already high level made possible by the great abundance of wood relative to the size of the population.

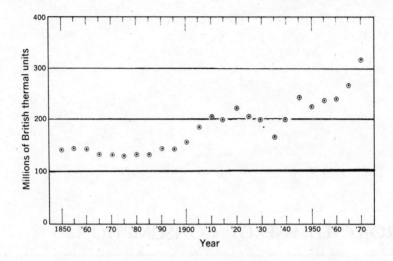

Figure 3.1.   Per capita consumption of energy in the United States, 1850 to 1970. Per capita consumption of energy has only a little more than doubled since 1850. Although the work output per capita has increased dramatically as heat engines substituted for work animals, the increase has been due to improved conversion efficiency without much change in per capita fuel input.

Growth in energy consumption still continues. Five key factors affecting the growth rate were identified and quantified in Chapter 1:

Factor	Annual Growth Rate (%)
Nonfarm working population	2.9
Energy content per unit of goods and services	0.9
Total population	1.3
Per capita residential basic energy	2.8
New energy consumed by all sectors	8.5

By estimating the future evolution of these growth rates, it is possible to project the future consumption of energy in the United States. Implicit in any such projection must be an assumption about the future availability and price of the basic energy sources. For many decades, with occasional fluctuations one way or the other, the prices of all basic energy sources have been declining gradually relative to the prices of other goods and services. The past growth rates of energy consumption must reflect this fact to some degree. I assume that this trend of the past will continue far into the future: that energy prices will continue their slow downward trend relative to other prices, with occasional fluctuations one way or the other. This assumption is carefully examined and justified in Chapters 5 and 8.

The nonagricultural working population cannot continue to grow more rapidly than the total population much longer. In 1968 agricultural workers amounted to only 5 percent of nonagricultural workers, down from 9 percent in 1960. If they all came off the farms, they would swell the ranks of nonagricultural workers only by the factor 1.05. At the same time about 40 percent of females over 16 years old were in the labor force, compared to 80 percent of males, with the female participation rate rising and the male participation rate falling, presumably because of more extended schooling, longer vacations, and earlier retirement. With increased daycare facilities for young children, combined with more care of infants by young fathers as the male labor force participation rate continues to decline, the participation rate for females could climb to that for males at something like 70 percent, giving a maximum increase in working population of a factor of 1.16.

The product of the factor 1.05 for taking the rest of the people off the farms and the factor 1.16 for taking the rest of the women out of the home is $(1.05)(1.16) = 1.22$, suggesting that the working population may grow another 22 percent relative to the total population. I assume that it will do so. This is enough to maintain the full $2.9 - 1.3 = 1.6$ percent differential growth rate between working population and total population experienced in 1960–1968 for another dozen years, or more likely to provide a diminishing differential growth rate for the balance of the century.

Total population growth rate is difficult to forecast, because it is so dependent on the shifting life styles of the population. I assume that the annual growth rate will decline slowly and uniformly from the 1.3 percent of the 1960's to zero in the twenty-third century, at which time the population will have stabilized at 1 billion people.

The energy content per unit of goods and services rose at the rate of 0.9 percent annually during the 1960's. Under the pressure of cost reduction, technology tends to decrease the energy consumed per unit of commercial and industrial activities. At the same time, an increased range and complexity of manufactured goods and an increased sophistication of commercial activities have tended to push more strongly in the opposite direction. I expect this trend to continue for some time. Examples include pollution control equipment for automobiles and powerplants, safety equipment for automobiles, heat pumps instead of simple furnaces for residential space heating, synthetic polymers instead of wood and natural fiber, refined coal instead of crude coal for fuel, and satellites instead of ground installations for weather observation and communication, all of which increase the energy content per unit of goods and services. Yet overall, I expect that cost reduction will gradually pull level with growth in complexity and sophistication of goods and services, so that the 0.9 percent will gradually decline to zero over the next 40 years.

Per capita residential consumption of basic energy grew very rapidly during the 1960's, at an average annual rate of 2.8 percent. Most of the increased energy use was for space and water heating. These are uses for which the average person can have enough—when the house is warm and the shower runs hot. I believe the 2.8 percent annual increase has been a result of increasing affluence, in which a growing proportion of the population is coming to enjoy the thermal comforts of home. There is evidence in the environmental movement that this trend is nearing its end. A few decades ago, we used to hear mainly the cries of the poor that they wanted more, but now we increasingly hear the cries of the affluent that they have enough. I believe them, and assume that in another 20 years nearly everybody will have enough basic residential energy, with the 2.8 percent declining to zero by the end of the 1980's. This amounts to about a 30 percent increase in per capita residential consumption of basic energy when this segment of energy utilization finally stabilizes.

New energy uses with their average 8.5 percent annual growth rate during the 1960's are growing at 7.2 percent annually on a per capita basis, largely as a result of increased saturation of lighting, television, air conditioning, and appliances. Here new technology has provided opportunities for novel forms of energy consumption, and increasing affluence is enabling a larger percentage of the population to take advantage of them. The future of new energy uses will be determined by the ability of technology to provide still further novel opportunities for energy consumption and by the level of affluence available for participating in them. Here I believe that technology has nearly run its course, and significant new uses for energy will not be forthcoming. I expect growth of this category to be largely the result of increasing affluence. (Note that an electric automobile, if ever economically attractive and socially acceptable, will represent neither new technology nor new energy use. Note that widespread use of heat pumps for space heating will represent neither new technology nor new energy use. Note that coal refining to produce synthetic crude oil or gas, if ever economically attractive and socially acceptable, is already included in the commercial, industrial, and transportation sectors as part of the gradual increase in energy content of goods and services. Note that increasing use of satellites for communication is also included there.) It is difficult to estimate the average degree of saturation of the various individual new uses of energy, but I assume that it is now 50 percent on the basis of the sketchy information available. This means that the per capita consumption of energy for this category of use will eventually double, with the annual per capita percentage increase declining from 7.2 percent to zero over the next 20 years.

The level at which per capita energy consumption will ultimately stabilize is inherent in the foregoing discussion. With respect to the basic energy consumption of the commercial, industrial, and transportation sectors, a factor of 1.22 comes about from greater participation of farm workers and women in the nonfarm working population, and a factor of 1.19 from the growth in energy content per unit of goods and services. A factor of 1.30 on basic energy consumption in the residential sector comes about from increased affluence, and a factor of 2 on the energy consumption of new energy uses comes about from the same cause. In combination these produce an ultimate factor of 1.49 on per capita energy consumption in the United States. The computation is summarized in Table 3.1. Per capita energy consumption is calculated year by year in Appendix 3, showing the pace at which it is forecast to reach its ultimate stable value.

The projected leveling off of per capita energy consumption is shown in Figure 3.2. With the assumed slow decline in annual population growth, the annual energy consumption of the United States is projected to double in the 30 years between 1970 and 2000, to double again in the following 70 years ending 2070, and finally in the 2200's to stabilize at about 7 times the 1970 level. The total quantity of energy to be consumed between 1970 and 2000 will

TABLE 3.1 PROJECTED SATURATION OF UNITED STATES PER CAPITA ENERGY CONSUMPTION

	1968 Fraction	Working Population Factor	Unit Energy Content Factor	Affluence Factor	Ultimate (relative to 1968)
Basic energy consumption (commercial, industrial, transportation)	0.737	1.22	1.19	1	1.070
Basic energy consumption (residential)	0.157	1	1	1.30	0.204
New energy consumption (all sectors)	0.106	1	1	2	0.212
Total energy consumption	1.000				1.49

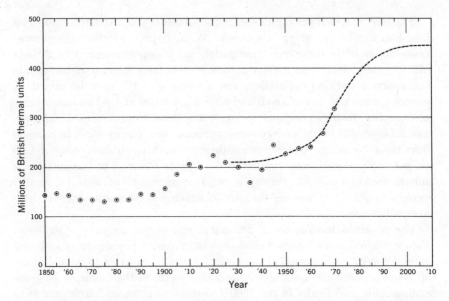

Figure 3.2. Per capita energy consumption in the United States, 1850 to 1970 and projected to 2010. Annual energy consumption is projected to level off at about 450 million Btu per person by the end of the century.

be about 300 C, and in the hundred years between 1970 and 2070, it will be about 1600 C.

For the other industrialized regions of the world, I assume that per capita consumption of energy rises to a saturation level about two-thirds of that for the United States, and that their populations grow only one-third as rapidly. Except for the Soviet Union and Oceana, the other industrialized regions have population densities nearly 10 times that of the United States, and they may be more nearly approaching population stability.

TABLE 3.2 PROJECTED ENERGY CONSUMPTION
(Units of C = 10^{16} Btu)

Region	30-Year Total 1970–2000	100-Year Total 1970–2070	Quality of Projection
United States	300	1600	highest
Other industrialized	800	2800	lower
Nonindustrialized	200	800	poor
World total	1300	5000	intermediate

It is much more difficult to project the future energy consumption of the nonindustrialized regions of the world. On a per capita percentage basis, their consumption of fossil fuels and hydroelectricity grew at about the same rate as that of the industrialized regions. On the other hand, their consumption of total energy (including traditional sources) grew substantially less rapidly than that of the industrialized regions. Some analysts imagine that per capita consumption in the nonindustrialized regions will rise to equal that of the industrialized regions. Others imagine that consumption of energy in the nonindustrialized regions may be highly concentrated in cities, with something like 10 percent of the population approaching the per capita energy consumption level of the industrialized regions, but the other 90 percent staying near the traditional level. This would mean that on the average the saturation level in nonindustrialized regions would be only about 10 percent per capita of that in the industrialized regions.

Because the nonindustrialized regions are proceeding with their growth in energy consumption from so small a base, and because their energy consumption growth rate is about the same as that of the industrialized regions, the two extreme assumptions about their mode of growth both lead to more or less the same result: per capita and total energy consumption for the nonindustrialized regions trailing far behind the industrialized regions for a century or more. Hence for the next century, I assume simply that the nonindustrialized regions' proportion of world energy consumption remains unchanged. Over this time span the percentage error for energy consumption in the nonindustrialized regions is likely to be large, but the resulting percentage error in world energy consumption is likely to be small.

In summary, the projected energy consumption for the next century, including fossil fuels, hydropower, and nuclear energy but omitting traditional energy sources of particular importance to nonindustrialized regions, is as shown in Table 3.2.

Next we turn to a consideration of the availability of energy reserves and resources, to assess their adequacy for supporting energy consumption at these levels.

CHAPTER FOUR

ENERGY RESOURCES

In attempting to determine the adequacy or inadequacy of the world's energy resources, the first step is to make as good an estimate as possible of the available flux of solar energy and of the fossil fuels, nuclear fuels, and geothermal heat reservoirs in the ground. The flux of solar energy is easiest to estimate, because it comes down from the sky and has been fairly well measured. Hydropower, as a special aspect of solar energy, is equally easy to estimate. The subterranean energy resources present more difficult problems, because they are usually hidden from direct view until reached by drilling and many of the promising areas of the earth have not yet been drilled.

The quantity of solar energy reaching the earth at the upper boundary of the atmosphere is well known, amounting to 530,000 C per year. But not all of this energy reaches the ground where it becomes available to man. About a third is reflected back into space by clouds, and some is absorbed by atmospheric constitutents. On average about half of the solar radiation reaches ground level, but some regions, such as the Middle East which is near the equator and free of clouds, get more than their share; and other areas, such as Alaska which is far from the equator and has many clouds, get much less. As rough rules of thumb, the highest solar energy flux at ground level, found in desert lands near the equator, can be taken as about 1 million Btu per square foot each year. The average flux over the lower 48 United States can be taken as about 0.5 million Btu per square foot each year. If the total energy supply of the United States in 1970, amounting to 6.45 C, were to be obtained from solar energy with an overall conversion efficiency of 10 percent, an area of about 50,000 square miles would be required. This area seems large, but it is only about 3 percent of the land now devoted to farms. It is clear that solar energy resources are far more than adequate for supplying the world with energy, although of course they may not be the least expensive.

Hydropower resources are limited to rivers and streams wherever they happen to be or wherever they can be diverted to. The total resource base—if all potentially developable falling water were turned to hydroelectricity—amounts to about 1 C annually in North America and 9 C annually in the world (4.1). This is not enough to make up the United States or world totals for energy, but is enough to allow hydroelectricity to grow for some time at its historic rate, amounting to 4 percent of United States energy input and 6 percent of world energy input (on a fuel-equivalent basis).

Next let us consider the various mineral fuels of current significance—coal, petroleum, natural gas, uranium, and thorium; and the mineral fuels of potential future significance—oil shale and tar sand. These minerals are not distributed uniformly throughout the earth, but are concentrated to varying degrees in deposits of varying size at different depths in different regions of the world. At any given time some of the known mineral deposits have been well mapped and have been developed for production, whereas others remain undeveloped. A deposit is developed when the necessary investment has been made in whatever capital equipment is required to produce the mineral from the deposit and to move it to market. For coal this means acquisition of mining rights, investment in mining equipment, establishment of the mine, and installation of transportation equipment to move the coal to the nearest existing railroad or waterway. For oil it means acquisition of oil rights, drilling of wells, installation of gathering lines and pumping equipment to move the oil to the nearest existing common carrier pipeline or shipping dock. Investments of this nature are made in anticipation and expectation of profitable production. The minerals that have been developed for profitable production are called reserves. (This definition of reserves is followed most closely in the oil, gas, and uranium industries, and perhaps less closely in the coal industry where well-known, easily developable deposits may be counted as reserves.)

Reserves amount to current inventory of minerals in the ground. As mineral production proceeds, material is withdrawn from inventory and reserves are diminished. At the same time, as investments are made in developing additional deposits, new inventory is created and reserves are increased. Current reserves at any given time reflect the interplay of these opposing tendencies. Reserves can be measured in terms of the reserve-to-annual-production ratio, which equals the number of years that the inventory would last if production continued at its present rate (and if no new inventory were developed). Economic forces keep the reserve-to-production ratio for many minerals at something like 10–20 years. Larger ratios tend to be uneconomic because the investment required to develop additional reserves would be made too long in advance of any expected return on the additional reserves. Small ratios tend to be uneconomic because a substantial portion of the useful life of the recovery equipment would be wasted if the inventory were used up too soon. Stated another way, reserves are developed to the point where the investment in marginal reserves will achieve a minimum acceptable return at the anticipated sales price. Should the reserve-to-production ratio for some mineral drop from 12 years supply to 10 years supply over some period of time, we need not necessarily assume that ultimate mineral depletion is approaching. It may reflect only prudent trimming of inventory at a time of rising interest rates, or perhaps it may reflect growing uncertainty about a potential flood of low-cost imports.

Known undeveloped deposits are often called submarginal, because they cannot be developed to produce minerals at a profit with today's technology, today's costs, and today's prices. Yet as the economy of scale reduces costs, and as new technology reduces costs, submarginal deposits tend progressively to be developed and are transformed into reserves. As an example, when oil is withdrawn from an oilfield, it flows easily at first, then less easily, then must be pumped, and finally, with whatever state of technology exists at the time, the cost of getting it out exceeds the price that it will bring. The inventory of recoverable oil has been exhausted. The reserves are gone. Yet 70 percent or so of the original oil in place in the field still remains there as a submarginal resource. As new recovery technology is developed, a time usually comes when it pays to redevelop the same field for additional production by means of water-flood or fire-sweep or some other new technology. New reserves are created whenever an old field is redeveloped. In principle, the process can be continued time after time until the oil is all developed. Similarly, when coal is removed from an underground mine, about half of the coal is left behind as pillars to hold up the ground above and prevent the mine from collapsing. This half of the coal is not counted as reserves, because the pillars cannot be removed profitably with today's cost, technology, and price structure. But the pillars remain as submarginal resources, along with undeveloped thinner and deeper seams, available for some future time when costs, technology, or prices may change.

In addition to reserves and known submarginal deposits, there are additional deposits not yet discovered. And (particularly for oil and gas) there is substantial worldwide activity devoted to discovering them. Once discovered, some will prove to be easily developable for low cost production, and these may be classified as unknown deposits of economically recoverable minerals. Others will prove not to be profitably developable, and these may be classified as unknown submarginal deposits. The quantities of undiscovered resources clearly are the most difficult to determine, but geological and geophysical experts have learned to make respectable estimates.

A simple diagram can be constructed for classifying the various types of mineral deposits as shown in Figure 4.1. A large rectangle is divided into four

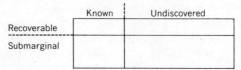

Figure 4.1.

smaller rectangles by vertical and horizontal dividing lines. The two small rectangles to the left of the vertical dividing line represent known deposits, and the two to the right represent unknown deposits. The two small rectangles above the horizontal dividing line represent recoverable minerals, and the two below represent submarginal minerals. The "known-recoverable" rectangle represents reserves, and everything else is called resources, as shown in Figure 4.2. At any given time the current state of knowledge of a mineral industry can

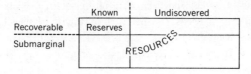

Figure 4.2.

be reflected by writing in each small rectangle a number reflecting its estimated magnitude and by drawing the rectangles so that their sizes are approximately proportional to the numbers.

U.S. Geological Survey specalists have compiled estimates of United States fossil fuel resources including coal, petroleum liquids, natural gas, and oil from shale (4.2). Their estimates are generally made on geological projections of favorable rocks and on anticipated frequency of the energy resource in the favorable rocks. The estimates of submarginal resources of oil from shale include only relatively rich deposits that might be recoverable with today's technology at less than 2 or 3 times today's oil prices, and they exclude much larger quantities of lower grade shale. These resource estimates were combined with authoritative reserve estimates in a series of scaled diagrams of the type just discussed. I believe that these U.S. Geological Survey estimates are the best objective estimates available for the United States.

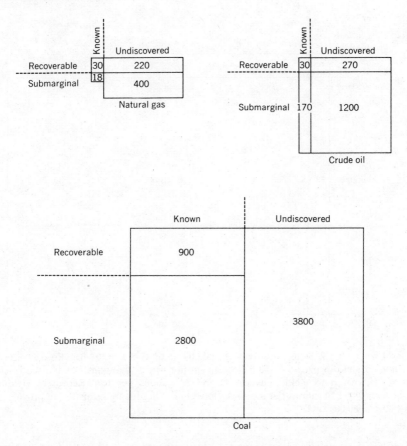

Figure 4.3.   United States reserves and resources of natural gas, crude oil, and coal, in units of C = 10^{16} Btu. These diagrams are all drawn to the same scale, as are others to follow in this series, to facilitate visual comparison.

Figures 4.3 and 4.4 are based on the Geological Survey scaled diagrams for fossil fuels (4.2). Figure 4.3 shows reserves and resources for natural gas, crude oil, and coal, the three fossil fuels currently being utilized in the United States. Figure 4.4 shows shale oil resources estimated to yield in excess of 10 gallons of oil per ton of shale. The figures are all drawn to the same scale, as are others to follow, to facilitate visual comparison.

Total cumulative United States consumption of fossil fuels has amounted to about 200 C through 1970, and 1970 consumption amounted to about 6.5 C. These quantities are drawn to the same scale in Figure 4.5 to show their magnitude in comparison with reserves and resources.

The data in Figures 4.3 and 4.4 show that the United States has abundant

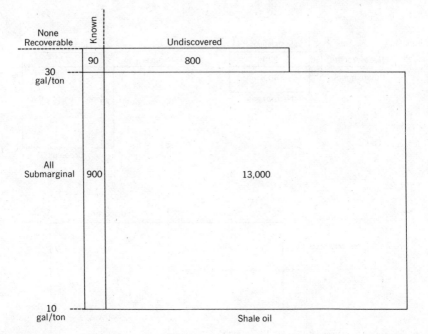

Figure 4.4.   United States resources of shale oil, in units of $C = 10^{16}$ Btu. Resources estimated to yield in excess of 30 gallons of oil per ton of shale are distinguished from
those estimated to yield between 10 and 30 gallons per ton. Resources yielding
between 5 and 10 gallons per ton, not shown in the figure, are estimated to provide an
additional 80,000 C (4.3).

fossil fuel reserves and resources, as summarized in Table 4.1. Present fossil
fuel reserves are more than adequate to supply the 300 C of energy required
for the balance of this century. If we accept the projections that annual per
capita energy consumption will level off at 450 million Btu per capita, as estimated in Chapter 3, and that annual per capita fossil fuel consumption for
nonenergy purposes such as asphalt tiles, road oil, and petrochemical feed-

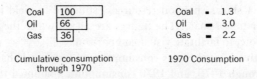

Figure 4.5.   Consumption of fossil fuels in the United States, cumulative through 1970
and for the year 1970, in units of $C = 10^{16}$ Btu. This figure is drawn to the same scale as
Figures 4.3 and 4.4.

TABLE 4.1 UNITED STATES FOSSIL FUEL RESERVES AND RESOURCES

Fossil Fuel	Reserves	Additional Resources
Coal	900	6,600
Petroleum	30	1,640
Natural gas	30	640
Shale oil	—	15,000
	960	24,000

stocks will level off at 50 million Btu per year (4.4); then, in an economy totally energized by fossil fuel, the annual per capita fossil fuel consumption would level off at 500 million Btu. If we assume in addition that the United States population will stabilize at about a billion people within the next few centuries, it follows that annual fossil fuel consumption (energy uses plus nonenergy uses) will level off at about 50 C per year. Under these conditions, in an economy in which all energy is derived from fossil fuels, United States fossil fuel reserves and resources would be enough to last about 500 years.

$$\frac{25,000 \text{ C reserves and resources}}{50 \text{ C per year}} \approx 500 \text{ years.}$$

Uranium and thorium contain so much energy per pound that it makes good sense to consider very low grade ores in estimating resources. The most comprehensive estimates have been made by an Interdepartmental Study (4.3) with participation by nine Federal departments and agencies. Their estimates are shown in Figures 4.6 and 4.7. The "current technology" energy values shown for uranium reserves correspond to utilization of 1.5 percent of the potential energy of the uranium, as is appropriate for today's light water reactors. "Breeder technology" energy values for both uranium and thorium correspond to 80 percent of the potential energy as may become possible with the new technologies of breeder reactors. Although nuclear fuel reserves available with current technology are not particularly large in relationship to projected energy consumption, nuclear fuel resources are very much larger.

Geothermal heat resources have been estimated by the U.S. Geological Survey (4.2) and are shown in Figure 4.8. Like fossil fuel, geothermal heat is a nonrenewable resource (4.5).

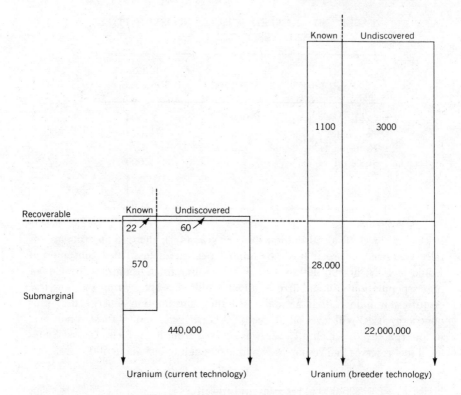

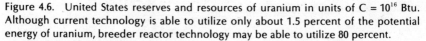

Figure 4.6.   United States reserves and resources of uranium in units of $C = 10^{16}$ Btu. Although current technology is able to utilize only about 1.5 percent of the potential energy of uranium, breeder reactor technology may be able to utilize 80 percent.

So far in this chapter consideration of fossil fuels, nuclear fuels, and geothermal heat has focused on the United States. The Interdepartmental Study (4.3) considered world resources as well as United States resources. Using data available in 1962, they estimated that in terms of total resources "the United States is endowed with approximately one-fourth of the coal, one-seventh of the oil, possibly one-tenth of the natural gas, one-twelfth of shale oil, and one seventeenth of uranium and thorium." Since the United States has about one seventeenth of the world's land area, it appears that it may have somewhat more than its share of fossil fuels, particularly coal. I expect that as better data become available, the energy resources of the world will prove to be more or less uniformly distributed. Most major countries should have their proportionate share, with a few exceptions such as Japan having none, the Mideast having more than its share of oil, and the United States having more than its share of coal.

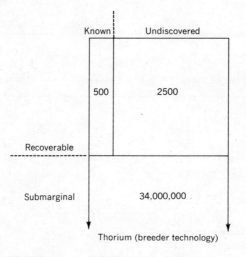

Figure 4.7.   United States reserves and resources of thorium in units of C = 10^{16} Btu, based on 80 percent conversion of thorium to fissionable nuclear fuels.

In summary, estimates of United States fossil fuel and nuclear fuel reserves and resources are as shown in Table 4.2. Fossil fuel reserves alone are adequate for 50 years. Fossil fuel reserves and resources alone are adequate for 500 years, and nuclear resources for a million years, provided, of course, that these reserves and resources can be utilized within the bounds of environmental acceptability. Solar energy is even more abundant. The energy reaching ground level in the United States amounts to about 10,000 C annually and is assured indefinitely. World energy resources are substantially

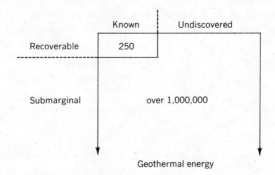

Figure 4.8.   United States geothermal heat reserves and resources in units of C = 10^{16} Btu.

TABLE 4.2

Energy Source	United States Reserves	United States Additional Resources
Fossil fuels		
Coal	900	6,600
Petroleum	30	1,640
Natural gas	30	640
Shale oil	—	15,000
Total	960	24,000
Nuclear fuels with breeder technology		
Uranium	1,100	22,000,000
Thorium	500	34,000,000
Total	1,600	56,000,000

larger, in approximate ratio of the world's land area to the United States land area.

Although it is true that nuclear and solar resources exceed fossil fuel resources by a wide margin, there is no immediate threat of fossil fuel depletion to cause us to shift to nuclear fuel or solar energy. The United States' supply of coal and shale oil is sufficient to last for many centuries if we choose to use it, and the same is true for the rest of the world. It is the overall socioeconomics of energy supply and utilization, to be discussed in later chapters, that will determine which fuels are actually used, and on what timetable.

CHAPTER FIVE

ECONOMIES OF SCALE

AND NEW TECHNOLOGY

In previous chapters we have looked at the trend of energy consumption in the United States, have dissected it into components, considered the factors that affect each, projected each into the next century, then added them all back together for a projection of total United States energy consumption. This projection assumed implicitly that long-term availability and price trends of fossil fuels would continue their historic trajectories. Review of reserves and resources showed that adequate quantities of fossil fuels were available, but the question of future prices was not addressed. This is a question that needs to be examined very carefully. We need to look broadly and deeply into the fundamental historic basis for prices and price movement. Some of the factors bearing on prices are based in straightforward economics, but some are primarily noneconomic. Both must be considered.

First let us consider a few basic economic factors. One of the most fundamental is the economy of scale, which I illustrate through the example of crude oil transportation by ocean tanker. A tanker is in effect a big bathtub shaped sheetmetal container, motorized for propulsion, with a crew to manage loading, unloading, and navigation. The costs of transportation are partly fixed charges associated with the capital invested in building the ship and partly charges associated with operating it.

Think of two tankers, a small one and a large one, built at the same time with the same technology, the small one to carry 100,000 barrels of oil and the large one to carry 800,000 barrels of oil. The larger ship carries 8 times as much oil, but it is only twice as long, twice as wide, and twice as deep. It has only 4 times the area of hull and deck. Hence if the hull and deck plates were of the same thickness of steel for each ship, only 4 times as much steel would be required for the large ship as for the small one. This would be only half as much steel per barrel carried. In practice, good ship design increases plate thickness in the large ship to make it structurally as sound as the small ship, but the required thickness increase proves to be less than a factor of 2, and there is still a substantial saving in the cost of steel per barrel of ship capacity. There is also a saving in the cost of propulsion equipment per barrel carried in the large ship, partly because a big propulsion system weighs and costs less per horsepower and partly because the large ship needs only about 3 times the power to go as fast as the small one. The costs of hull and propulsion equipment are the major capital costs of ship construction. Per barrel of capacity, they are smaller for the larger ship; hence the larger ship benefits from an economy of scale in ship construction.

Operating expenses for the large ship also are less than those for the small ship, per barrel carried. Since the propulsive power per barrel is less for the large ship, less fuel is consumed per barrel transported. The same size crew can handle either ship, since the type and number of sea duties to be done are not dependent on ship size, leading to an eightfold saving in crew costs per barrel for the larger ship. Hull and deck maintenance of the large ship consumes only half as much manpower and paint per barrel of ship capacity.

In aggregate, capital cost and operating cost factors combine to produce an approximately threefold overall reduction in total transportation cost per barrel for the larger ship (5.1). This amounts approximately to a 30 percent cost reduction for each doubling of tanker capacity, a significant economy of scale.

Most ocean tankers now under construction have capacities in the neighborhood of 250,000 tons, with the largest close to 500,000 tons. Yet nothing prevents the construction of even larger tankers with even lower unit costs. History suggests that each of the largest tankers carries about a third of 1 percent of the world's ocean oil traffic and that tanker sizes increase as oil traffic increases. If world oil traffic doubles again, as it may well do, we can anticipate that the size of the largest tankers will double too, and that the cost of transporting a barrel of oil will decline another 30 percent.

As a second example of the economy of scale, consider coal mining. The capital costs of mining equipment and the operating costs of mining both decline, per ton of coal mined, as the daily mine output is increased. The reasons for this economy of scale are very similar to those for tankers. Equipment with doubled mining capacity weighs less than twice as much; hence it costs less than twice as much, and the equipment cost per unit capacity is reduced. Operating costs are reduced as well because fewer operators are required per unit capacity. Studies of strip mining costs by the U.S. Bureau of Mines (5.2) show that the cost per ton of coal is reduced approximately 20 percent by each doubling of mine capacity.

Many of the activities that affect fuel costs respond to the economy of scale. Among the most significant are mining, pipeline transportation, tanker transportation, ore milling, and oil refining. Economies of scale lead to unit cost reductions whenever the output of the average unit of production (the capacity of the average tanker, the production of the average mine, the capacity of the average pipeline, the output of the average refinery) increases with the passage of time. This usually occurs in a growing industry, but it can occur through consolidation of production facilities in a static or declining industry.

Another economic factor of significance, similar in effect to the economy of scale, is the stream of cost reductions and performance improvements made possible by new technology, spurred by competition. I call it the economy of new technology. As an example, consider the influence of new technology on the performance of computer processors. As transistors were improved, the speed of data processing could be increased without changing the physical size of the processor at all. Output was increased much more rapidly than cost, and the unit cost of a given computer operation declined. Similar examples can be found in energy-related activities. The discovery by ship designers that a subsurface bulblike nose on the front end of a tanker would increase tanker speed at very little cost enabled more barrels to be transported by a given ship each year and decreased transportation costs per barrel. When gas turbine designers improved the aerodynamics of compressor blades, they increased the power output and efficiency of a given turbine at relatively little cost and decreased both capital and operating costs per horsepower.

Very often the economies of scale and new technology proceed concurrently, and it is difficult to separate them. The overall effect is to cause unit costs to decline steadily—the more units that are manufactured, transported, or processed, the lower the unit cost. The unit cost reduction that accompanies an increase in production often approximately follows a simple mathematical relationship that can be stated as follows:

> Every time there is a doubling of the total cumulative number of units manufactured, transported, or processed in some economic activity, there results a fixed percentage decline in unit cost.

The curve of declining cost versus cumulative industrial activity is sometimes called a learning curve or experience curve to highlight the role of the economy of new technology, although economy of scale is of greater significance for many activities. The fixed percentage of decline depends on the nature of the economic activity and is often somewhere near 20 percent.

Since inflation and deflation change the value of money from year to year, it is necessary to correct for them if the true picture of cost reduction through increased experience is to be revealed. I use the overall GNP deflator described and tabulated in Appendix 4 to adjust dollar figures to 1970 dollars. All costs or prices in this book have been converted to 1970 dollars. I mention this fact from time to time as a reminder, but it is true whether mentioned or not.

Except for raw fuel recovery at mines and wells where progressive depletion is a complicating factor, most energy-related activities, including transportation, refining, electric power generation, and distribution of refined products, would be expected to follow normal experience curves with steadily declining unit costs. It is useful to develop experience curves for some of these activities and thereby to establish a basis for projecting future energy costs as experience continues to accumulate. Two factors need to be ascertained: cumulative production and unit cost, both over as long a period of time as is practicable. Cumulative production figures are readily available, but unit costs are not. The best that can be done is to utilize unit prices in the hope that price will measure marginal cost (including a minimal return).

The hope that price will equate to marginal cost may be justified where competition is vigorous. This tends to occur at the point of sale of a refined energy product that embodies the combined costs of crude fuel, transportation, refining, and distribution of refined products. Each component of the total enterprise tends to be adjusted so that the total overall cost is minimized, and there is ample opportunity for economies of scale and of new technology to be applied all along the line. Gasoline at the retail pump is such a product, and I assume that its price reflects its aggregate cost. Because the cost of crude oil is affected by depletion and is not expected to follow a normal experience curve, it is desirable to subtract it out and treat it separately. The remaining cost of retail gasoline processing, including transportation, refining, and distribution, is expected to follow a normal experience curve since it is unaffected by depletion. Figure 5.1 shows the resulting experience curve for retail gasoline processing. The curve is approximately linear on a log-log (double ratio) scale, as an ideal experience curve should be, and shows that the cost of gasoline processing (measured in 1970 dollars) declined by approximately 20 percent, to 80 percent of its former value, every time cumulative production doubled.

Turning now to the cost of crude oil, Figure 5.2 shows the average wellhead price of crude oil in the United States for the past century, as it depends on cumulative production. Through more than eleven doublings of cumulative production, crude oil price has declined slowly, with rather large fluctuations, by roughly 5 percent of its previous value for each doubling. Although oil prices have been subject to manipulation and have not always been determined by free competition, it is still possible to hope that the overall trend in price reflects an overall trend in marginal cost, and that Figure 5.2 represents a slow decline in the cost of crude oil with the accumulation of experience. One reason for the relatively slow decline in crude oil cost per doubling of cumulative production is the fact that crude oil recovery does not respond to the economy of scale. When the rate of oil recovery is doubled, experience and analysis both show that the capacity of the average well stays about the same and that the number of wells is doubled. Only the economy of new technology remains as a significant factor in crude oil cost reduction.

The cost of retail gasoline is dependent on the costs of both gasoline processing and crude oil recovery, and the future cost of retail gasoline can be projected by projecting these two cost components and combining them (5.3). The future cost of processing can be estimated simply by projecting the experience curve in Figure 5.1, for there is no barrier to prevent the economies of scale and new technology from continuing to reduce costs as additional experience accumulates. The future cost of crude oil recovery is a different matter because of the progressive effects of depletion.

Crude oil recovery represents a very special type of industrial activity for which progressive depletion introduces a significant additional cost factor beyond the economies of scale and new technology. Mineral recovery costs in general tend to rise, because recovery tends first to exhaust rich deposits, then to move on to leaner deposits with higher unit costs. The overall unit cost of a mineral at the mine or well can either rise or fall depending on the relative significance of the economies of scale and new technology, which work to lower

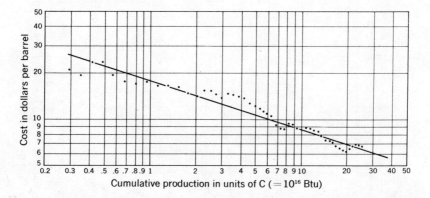

Figure 5.1. Experience curve for retail gasoline processing, for the United States 1919 to 1969. Gasoline processing begins with crude oil at the well and includes all subsequent activities such as transportation, refining and distribution (5.3). Cost is measured in 1970 dollars. The trend line corresponds to a 20 percent decline in processing cost for each doubling of cumulative production.

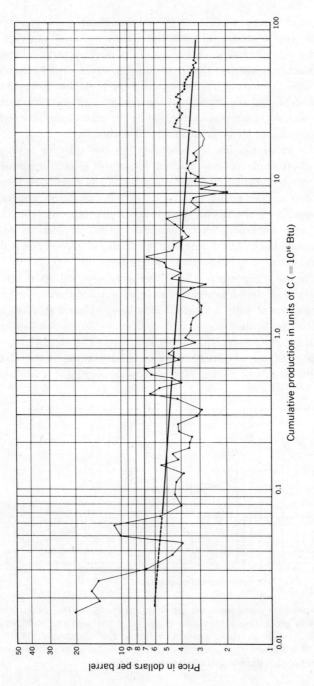

Figure 5.2. Experience curve for crude oil at the well, for the United States 1869 to 1971 (5.4). Price is measured in 1970 dollars and is assumed to approximate marginal cost (including a minimal return). The trend line corresponds to a 5 percent decline in cost for each doubling of cumulative production.

unit costs, and the progressive use of leaner and leaner deposits as richer ones are depleted, which works to raise costs. Yet all cost factors including depletion are dependent on cumulative production, and I assume that unit costs associated with mineral recovery can be described by a cost-versus-experience relationship suitably modified to account for depletion. Specifically, I assume that the usual rule of experience can be modified for mineral recovery by introducing a multiplying factor that increases cost with increasing depletion. The rule of experience becomes a two-part rule, the second part applicable only to mineral recovery:

A. Every time there is a doubling of the total cumulative number of units manufactured, transported, or processed in some economic activity, there results a fixed percentage decline in unit cost.

B. For mineral recovery, where f stands for the fraction of the original total resource base that remains in the ground, the unit cost in part A is multiplied by the factor $1/f$.

Part A of this rule has been tested and found applicable to many industrial activities, but part B is only a working hypothesis. We do not have sufficient experience to know how rapidly mineral recovery costs tend to rise as ultimate depletion is approached. Yet this factor is so important for anticipating future mineral recovery costs that a tentative estimate is desirable for illustrative analysis.

Now it is possible to apply the modified rule of experience—parts A and B combined—to the recovery of crude oil and to a projection of the cost of crude oil recovery. The straight line drawn in Figure 5.2 corresponds to part A with a 5 percent reduction in cost for each doubling of experience. To take account of depletion, the cost value at each point on the line must be multiplied by the appropriate depletion factor $1/f$. In 1970, for example, cumulative oil production since the beginning of the industry amounted to about 3 percent of the original resource base. The fraction remaining in the ground was $f = 0.97$, and the depletion cost factor was $1/f = 1.03$. Hence the cost value corresponding to 1970 on the crude oil experience curve must be multiplied by 1.03—that is, raised by 3 percent—to take account of part B of the rule of experience. When the entire experience curve in Figure 5.2 is modified to take account of part B of the rule of experience, the modification is hardly noticeable, suggesting that depletion has not been a significant factor in the cost of crude oil recovered up to the present time. But the situation changes when the cost of crude oil is projected into the future, as is done in Figure 5.3.

The projection in Figure 5.3 suggests that the cost of crude oil is nearing its all-time minimum, which may occur in a decade or two, at about one more

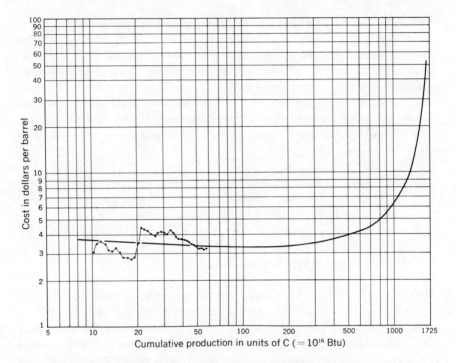

Figure 5.3.   Cost of United States crude oil at the well 1935 to 1941 repeated from Figure 5.2, and projected toward depletion for a total resource base of 1725 C.

doubling of cumulative production, and that crude oil cost will thereafter rise progressively as ultimate depletion is approached. [The minimum is very flat, however, and the cost of crude oil is projected to vary less than 5 percent over two or three doublings of cumulative production centered around it. Although the precise position of the crude oil cost minimum is dependent on the form of the depletion cost factor, assumed to be $1/f$ in this illustrative analysis, a minimum is projected to occur within the next one or two doublings for any plausible form of the depletion cost factor. Much the same history can be expected for natural gas, whose minimum cost may already have been reached because the depletion of natural gas resources has reached about 6 percent compared with crude oil's 3 percent.]

It is important to keep in mind that natural crude oil is not the only potential source of refined products. The full spectrum of refined hydrocarbon products can be produced from synthetic crude oil derived from solid fossil fuels such as coal, oil shale, or tar sand. Germany produced synthetic crude oil and refined products, including aviation gasoline, motor gasoline, diesel oil, heating oil, and lubricants, from coal during World War II, and South Africa produces these products from coal today.

The cost of synthetic crude oil from coal was so high that Germany abandoned this process after the war when natural crude oil became available. The cost in South Africa presumably also is higher than the cost of natural crude oil, but coal is indigenous to South Africa whereas oil is not, and strategic considerations may justify the higher cost. Although the production of synthetic crude oil is not now economically attractive, there is every reason to believe that it will become so in the future. When it does, we may anticipate the establishment of a growing synthetic crude oil industry with the price of synthetic crude oil determining the price of all crude oil.

To see how this may come about, let us consider the general question of starting up a brand new industrial activity. If the designers' plans are correct, the first plant will be marginally profitable. A second plant should be more profitable because of the experience gained with the first. Since a second plant would double industry experience, it might be expected to lower unit costs to about 80 percent of those of the first plant. A fourth plant would double experience again with another 20 percent cost reduction, and an eighth would double it a third time with still another 20 percent cost reduction. After 10 plants the unit cost might be halved.

Proponents of processes for making synthetic crude oil from coal, shale, and tar sand are currently of the opinion that their processes would be competitive at production rates of about 100,000 barrels per day (5.5). By way of comparison, the largest German plant in 1944 was capable of producing about 10,000 barrels per day, and the entire German synthetic crude oil industry was capable of producing about 100,000 barrels per day (5.6). The economy of scale applies to both mining and refining of solid fuels, so that a 100,000 barrel per day plant should yield products at about half the unit cost of the German 10,000 barrel per day plant, which might indeed make the process competitive. Since the first 100,000 barrel per day plant can build on the German experience, and would just about equal the total German capability, we may approximate its position on an experience curve of new 100,000 barrel per day plants by counting it as number two instead of number one. This would mean that about 20 such plants would be required to halve the cost of synthetic crude oil (instead of 10). Twenty plants would produce 2 million barrels per day or about 10 percent of anticipated United States oil consumption in 1980. At this production rate the cost of synthetic oil would be perhaps $2 per barrrel (1970 dollars) compared with more than $3 per barrel in the absence of synthetic crude oil availability.

Clearly, if the proponents of synthetic crude oil are right in their cost estimates, and if a plant with 100,000 barrel per day capacity would be marginally profitable, a great synthetic oil industry can be expected to follow the first commercial plant that demonstrates the favorable economics. The inhibiting factor that prevents a rush to build synthetic crude oil plants today is the fact that new plant costs tend to be higher than estimated more often than lower. If the cost of synthetic crude oil from the first 100,000 barrel per day plant should turn out to be 50 percent higher than anticipated, the project would be a financial disaster, showing in an expensive way that a plant about 4 times larger would be required for marginal profitability.

Figure 5.4 shows an estimate of the experience curve for synthetic crude oil made from coal. It is based on the following two assumptions:

1. Aside from depletion, the cost of synthetic crude oil from coal will decline by 20 percent for every doubling in cumulative production, because coal mining and the other synthetic crude oil processing steps all respond to both the economy of scale and the economy of new technology. Because mining responds to the economy of scale (whereas crude oil recovery does not), the minimum cost of synthetic crude oil is not reached until about 25 percent depletion (compared with about 6 percent depletion for natural crude oil).

2. The present cost of synthetic crude oil is double the cost of natural crude oil, and corresponds to a cumulative production of 0.1 C (the total German production). This assumption is consistent with the idea that a 100,000 barrel per day plant would be marginally economic.

With these assumptions, the cost of synthetic crude oil from coal is projected to reach a minimum of about $0.40 per barrel (measured in 1970 dollars) at about 25 percent depletion of the coal resource base. The minimum is rather broad, and the cost does not rise to $1.00 per barrel until the coal is about 80 percent depleted. Hence most of the coal resource is available for low-cost synthetic crude oil production, and the same is true for oil shale.

The cost of refined products such as gasoline includes both processing cost and crude oil raw material cost. We have seen in Figure 5.1 how gasoline processing costs have declined with cumulative production, and I expect that they will continue to decline indefinitely along the same experience curve trajectory. The processing costs of other refined products are expected to decline similarly. The cost of crude oil in the future will depend in part on the timing and extent of synthetic crude oil production. The sooner and more rapidly synthetic crude oil facilities are brought into production, the more rapidly the cost of crude oil will fall. We may anticipate that, from time to time, demonstration synthetic crude oil plants will be built and the current eco-

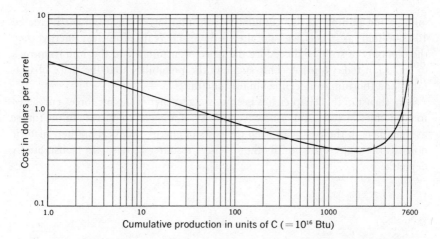

Figure 5.4. Hypothetical cost of synthetic crude oil manufactured from coal, based on assumptions stated in the text, projected toward depletion for a total resource base of 7600 C.

nomics will be tested. When production costs are favorable, a synthetic oil industry will take off. It is the practical certainty of this happening at whatever time in the future the price of natural crude oil begins to rise, if indeed the synthetic oil industry does not take off before then, that enables me to project with confidence a continued decline in the price of fluid hydrocarbons. I expect that economic and technological pressures will cause the shift from natural to synthetic crude oil to begin at just the right time to keep the refined products flowing on their historic downward (constant dollar) price trajectory.

The shale oil resources summarized in Chapter 4 considered only shale with recoverable oil content of 10 gallons or more per ton, a United States total of 14,500 C. Had the resource base been cut off at shale yielding 5 gallons or more of oil per ton, the resources would have amounted to 95,000 C, over 6 times as much (4.3). If we stay with fossil fuel that long, the economy of scale will enable us to produce even cheaper oil from this leaner shale. Hence the price of liquid hydrocarbon fuel—whether from refined oil now, refined coal later, or refined shale ultimately—could decline for millenia.

The situation with respect to uranium and thorium is similar. At present we are using ores with nuclear fuel concentrations of about 2000 parts per million. Much of the rock in the earth's crust contains about 10 parts per million of uranium and thorium, a concentration 200 times lower. If we were to mine that rock today to produce the same output of nuclear fuels that we produce today, we would have to mine 200 times as much rock to do it, yet the cost would be only about 35 times higher because of the economy of scale in mining and milling. If our cumulative consumption of nuclear fuels increased 100,000-fold, as it may well do after a thousand years of nuclear energy, then the economy of scale could reduce the price of uranium and thorium from rock to less than today's price. This suggests the possibility that the price of uranium and thorium can decline almost indefinitely.

The thrust of this chapter has been to suggest that economic forces embodied in economies of scale and of new technology, if allowed to operate in the future as they have in the past, can keep the costs of energy products on their historic downward trajectories until the energy resource bases are substantially depleted.

CHAPTER SIX

SOCIAL AND ENVIRONMENTAL

FACTORS

If economic factors such as the economy of scale, the economy of new technology, and the mineralization patterns of different fuels were the only factors of significance, and if competition among different fuels and fuel producers were to prevail, the projections in the previous chapter of universal price declines (in constant dollars) for refined fossil and nuclear fuels might be considered as price decline forecasts. But social and political factors have profound influences on many aspects of the world's energy activities, and it is prudent to weigh their potential impact as carefully as possible before promoting projections to the status of forecasts.

Two social factors of significance to the pattern of growth in energy consumption in the United States have been identified and analyzed in Chapters 1 and 3: first, the migration of workers from the country to the city, and hence from farms as workplaces to factories and offices as workplaces and second, the rapid growth in the proportion of women in the work force. These trends have caused both the number of workers per capita and income per capita to rise, and these effects have caused the consumption of energy to rise more rapidly than the population. In one sense the migration to the cities and the entry of women into the work force are economic factors, because they have economic consequences. Yet the motivations for these migrations are more complex than anticipation of potential economic consequences. They are based in addition on a mixture of human aspirations, envies, ideals, beliefs, convictions, and faiths. The civil rights movement and the women's liberation movement are based in the ethical concept of equal rights and equal opportunities as much as in the economic objectives of allowing blacks to leave the farms and women to leave the kitchens to enter the industrial and commercial work forces. These movements may be considered as social factors, since a significant part of their motivation lies outside economics.

The popular movement to limit family size, based in part on the availability of reliable oral contraceptives, and the movement wherein a portion of the population is adopting a low-energy-consumption life style, are examples of noneconomic social factors that may affect future energy consumption in a negative direction.

Concern for the environment, recently and unexpectedly aroused in the United States, is having profound influences on the pattern of energy utilization and may extend its influence to the rest of the industrialized world. I view the environmental movement as the social factor of primary importance to energy availability and utilization, and it merits discussion at length. Although there are elements of anti-establishment, antitechnology, and antigrowth sentiment behind the environmental movement, I believe that these motivations are secondary to the basic concern that we may be increasingly and excessively polluting and spoiling the air, water, and land we all share in common.

Every step of the way, from mine or well through the ultimate use by consumers, the energy industries affect the environment:

Industry Segment	Key Environmental Concerns
Fossil fuel recovery	Oil spills at offshore wells
	Strip mining unsightliness and acid drainage
	Underground mine safety
Fossil fuel transport	Ocean pollution by tankers
	Arctic pipelines
Fossil fuel refining	Air and water pollution
Fossil fuel combustion	Atmospheric pollution
	Thermal pollution of rivers and estuaries
Hydroelectric power	Inundation of land
	Silting of reservoirs
Nuclear power	Radioactive waste disposal
	Transport of radioactive materials
	Plant radioactivity levels
	Fear of accident
	Thermal pollution of rivers and estuaries
	Nuclear black market
High-voltage transmission	Aesthetics

In general these concerns reflect an increasing public awareness of undesirable environmental side-effects of energy utilization, and of equal importance, an increasing public awareness of our ability to substantially alleviate or eliminate such side-effects. In affluent societies, there tends to be a steadily diminishing level of tolerance for putting up with avoidable nuisances. It is highly probable that government regulatory power will be employed to solve the problems of pollution as it previously has been utilized to provide us with disease-free milk, pure food, and pure drugs. In the past, when it became possible to eliminate public nuisances or dangers, regulation forced industry to take the appropriate steps, and the public willingly paid the higher price required. Now it is the energy industry's turn, as current environmental legislation makes clear.

The Environmental Protection Agency has been established and given wide responsibility for interpretation and administration of antipollution legislation. Major control of atmospheric pollution from fossil fuel combustion is the goal of the Clean Air Act of 1970, which specifically limits the concentrations of sulfur oxides, particulates, nitrogen oxides, carbon monoxide, photochemical oxidants, and hydrocarbons in the air and in the products of combustion, and sets up a timetable for compliance. Other legislation is addressed to controlling pollution of the nation's waterways. The Atomic Energy Commission participates in setting radiation limits that must be satisfied by nuclear power plants and nuclear fuel reprocessing plants, controlling and limiting the radiation that can reach the public through the air and cooling water, or by any other route. Representatives of the environmental movement have not left the government regulatory agencies and the energy industries alone to work things out. They have intervened in regulatory proceedings to make sure their viewpoint was appreciated and have gone to the courts to make sure it was effective.

It is instructive to review some of the technological and economic changes already brought about by the environmental movement, and to anticipate what other ones may occur.

Because of the limitations on sulfur oxide emissions, many coal-burning electric power plants, particularly in the Northeast, found that they could no longer continue legally to burn the high-sulfur coal that was mined in their geographical area and that they had been burning for years. In principle they could have installed sulfur-removal equipment to cleanse their stack gases of sulfur, but reliable equipment of this type was not available, the need for such equipment having become apparent too late for technology to respond in time. They could have brought in low-sulfur coal from the western states, but the transportation cost would have been high, and the western mines were not ready to meet such stepped-up demand. Thus many of them chose to switch from coal to low-sulfur oil. The switch by northeast utilities increased the de-

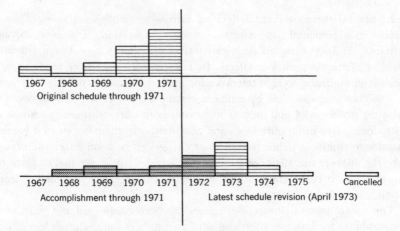

Figure 6.1.   Nuclear plant delays. Each block represents one of the 31 nuclear plants originally scheduled to go into commercial operation in the 5-year period 1967 to 1971. Of these 31 plants, 10 went into commercial operation within the originally scheduled 1967 to 1971 period, 6 went into commercial operation in 1972, and 1 in the first quarter of 1973. One plant was cancelled. As of April 1, 1973 the 13 remaining plants were rescheduled for the years 1973 to 1975 (6.1).

mand for petroleum and for refining capacity beyond that which had been anticipated by the petroleum industry who—along with most others in industry—were taken unaware by the environmental movement.

Construction of nuclear plants has been delayed, partly because of the usual growing pains of a new technology, but partly and significantly because of environmental legislation (much of it passed after plant design and construction had begun). In addition there have been intervention and legal action by environmental groups. Delays averaging several years have been experienced, as illustrated in Figure 6.1.

The upper portion of the figure relates to the situation at the beginning of 1967, when the first large nuclear power plants had been planned and scheduled. Many were under construction, but none was yet in commercial operation. Thirty-one nuclear plants were scheduled by the electric utility industry to be producing electricity by the end of 1971. The industry was confident of meeting this timetable because of their long history of bringing conventional power plants into commercial operation on schedule.

The lower portion of the figure shows accomplishment as of April 1, 1973. Of the 31 plants scheduled to be in commercial operation by the end of 1971, only 10 had made it. Six more had gone into commercial operation in 1972 and one in the first quarter of 1973. One plant had been cancelled. Thirteen plants had been rescheduled for 1973–1975. Unanticipated construction and licensing delays had ruined the 1967 schedule, the average plant being delayed about 2 years (neglecting the cancelled plant and assuming no further delay for the 13 plants not yet in commercial operation).

Two significant consequences arose from the nuclear plant delays illustrated in Figure 6.1. First, such delays are expensive and add to the cost of nuclear power, as is discussed in Chapter 13. Second, by the end of 1971 the utility industry found itself short 21 nuclear plants that it had been counting on for 16,-000 megawatts of generating capacity. (Meanwhile, other environmentalists were delaying the construction of pumped storage hydroelectric plants, so that the total deficit in generating capacity was somewhat greater, amounting to about 19,000 megawatts.)

It takes a long time, after the initial decision by an electric utility to add new nuclear capacity, before the new plant goes into commercial operation. Something like 12–13 years are required to go through all the steps from the initial search for a suitable site through site acquisition, plant design, construction, licensing, testing, and finally commercial operation. When the utilities found themselves with a generation capacity deficit because of nuclear and pumped storage delays, they had little choice of where to turn to get the generation capacity they needed to fill the gap. Only oil-burning gas turbines, similar in many respects to the turbines that power jet aircraft, could be manufactured and installed rapidly enough to meet the remaining very short timetable. The utilities turned to them in desperation, just managing to bridge the nuclear generation gap as shown in Figure 6.2, and the threat of widespread blackout was averted.

Although the utilities had not planned the introduction of gas turbines on anywhere near the scale made necessary by nuclear delays, it turned out to

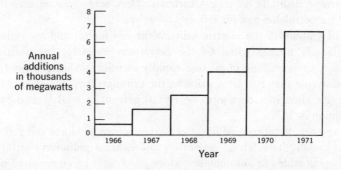

Figure 6.2. Annual additions to gas-turbine-powered electric generating capacity for the years 1966 to 1971 (6.1). About 21,000 megawatts of capacity had been added by the end of 1971, just in time to overcome the 19,000 megawatt deficit due to nuclear and pumped storage plant delays (see text and Figure 6.1).

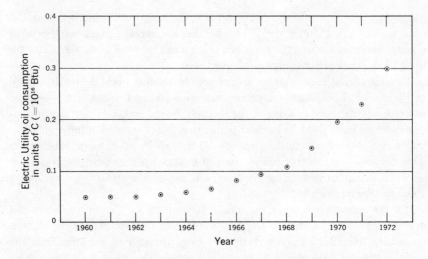

Figure 6.3. Consumption of oil by the electric utility industry increased sixfold in a decade, to nearly 10 percent of total United States oil consumption (6.2), largely in response to environmental regulations.

make good economic sense, as is discussed in Chapter 13. And the hothouse environment of the gas turbine manufacturing business forced rapid technological development of gas turbines, so that the machines available today are far more efficient and far less expensive than would otherwise have been the case. The gas turbine industry boomed as the nuclear industry reeled, and gas turbines have won a permanent role in the generation of electricity. The impact on the demand for fuels was equally dramatic. Instead of powering new capacity with uranium in nuclear plants as anticipated, the electric utilities were burning distillate oil in gas turbines. Here was a second unanticipated demand for petroleum and for refinery capacity.

The magnitude of the electric utility demands for oil and for refinery capacity, largely unanticipated by the petroleum industry, are illustrated in Figure 6.3. Consumption of oil rose rapidly as utilities shifted away from the high sulfur coal they had been using or the uranium they had expected to use, increasing sixfold in a decade to nearly 10 percent of total United States oil consumption in 1972.

Still another unanticipated demand for petroleum products and refining capacity is making itself felt as new safety and exhaust pollution control equipment is being added to automobiles. More gasoline is being required per mile driven, and more refinery capacity is being required to refine it.

In the coincidence of these various unanticipated demands for petroleum products and refinery capacity, we can identify one cause of the current energy crisis in the United States. These coincident demands were not forseen because the environmental movement was not forseen, and capacity was not provided to meet them. The demands could and would have been met if anticipated, and they can and will be met within two or three years as oil production and refinery capacity adjust to the new situation. In the meantime we can expect shortages and temporarily higher prices, as Europe experienced during the Suez Canal closure crises that stopped the flow of oil from the Middle East to Europe temporarily during 1957 and permanently in 1967. And we can also expect the shortages to disappear in a few years as Europe experienced after the Suez crises.

Yet permanent effects will remain, as permanent effects remained after the Suez crises. Four responses to Suez were to develop oil reserves in nearby North Africa, to discover oil and gas in the North Sea, to build a new generation of very large tankers that could economically transport oil around the African continent, and to speed the construction of nuclear power plants. These actions would all no doubt have been undertaken in due course under normal conditions, but the crises brought them on with a rush, and their effects did not fade away when the crises faded away. The strategic and economic importance of the canal has been diminished because large tankers cannot operate in its shallow water. They must continue around Africa whether or not the canal is reopened. The North African oil continued to flow, diminishing the importance of the Middle East. The North Sea exploration produced significant oil and gas discoveries and a boom of development and further discovery. The nuclear plants continue to operate independently of oil. Overall the dependence of Europe on Middle Eastern oil declined, and the energy supply became more secure and reliable.

Among the permanent effects of the environmental portion of the United States energy crisis of the early 1970's are likely to be found a vigorous gas turbine power plant industry, built to compensate for nuclear plant delays as already discussed, and a switch to higher-cost clean fossil fuels for electric power generation and other industrial uses.

At the end of the 1960's, electric power plants and other industrial plants were responsible for most of the sulfur oxides and particulate matter and much of the nitrogen oxides put into the atmosphere. Space heating and transportation had been converted years before from unrefined coal to refined petroleum and natural gas, so that electric generation and industry with their tall and visible smokestacks remained alone and exposed as major burners of dirty fuel and as major atmospheric polluters. There is little question that social pressure against atmospheric pollution will continue and that the shift from crude coal and high-sulfur oil to refined fuel begun on the east coast will sweep across the country in the 1970's. The refined fuel may be natural gas, refined

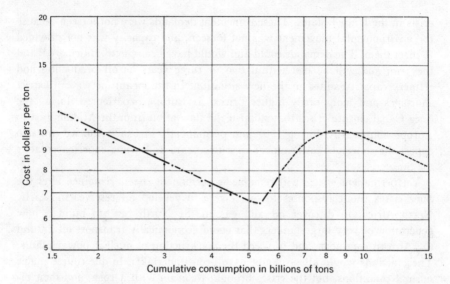

Cumulative consumption in billions of tons

Figure 6.4. Experience curve for United States electric utility coal, 1948 to 1971 (6.2) and projection. The trend line corresponds to a 25 percent decline in cost for each doubling of cumulative production. Cost (measured in 1970 dollars) followed the trend line closely for two decades through 1969. Then, on enactment of the Clean Air Act of 1970, cost rose substantially in 1970 and again in 1971 as utilities began to switch to coal with lower sulfur content. The projection assumes that the cost of coal will be doubled by the time the switch is complete.

petroleum or refined coal, but in any event it is likely to be considerably more expensive than the crude coal or high-sulfur oil it replaces. Utilities or factories that follow the alternative route of cleaning their stack effluent, which corresponds roughly to refining the fuel after it is burned, are likely to find that route equally expensive. The overall results will be cleaner air and more expensive fuel for electric generation and industry. It is important to keep in mind that this rise in cost of fossil fuel for electric generation and industry results primarily from the switch to clean fuel, which has always been high priced, away from dirty fuel, which has always been low priced, and not from price changes in individual fuels.

The shift to more expensive fuel began immediately after enactment of the Clean Air Act of 1970, as shown in Figure 6.4. For many years the cost of coal for electric power generation had been declining with cumulative production along a typical experience curve as the result of economies of scale and new technology in mining and transportation. Then in 1970 and again in 1971 the cost of coal jumped dramatically, reaching a value 30 percent above an extension of the old experience curve. The rising cost of coal represents the combined results of stricter enforcement of mine health and safety laws and of progressive substitution of high-delivered-cost coal with low sulfur content for low-delivered-cost coal with high sulfur content. When this substitution is complete, the cost of electric utility coal may rise as far as 100 percent above an extension of the old experience curve, after which the downward trend of a normal experience curve might be expected to resume, as projected in the figure. The higher cost of clean fossil fuel results in a higher cost of electric power, as discussed in Chapter 13.

The social factors that have been considered—the civil rights, women's liberation and environmental movements—do not appear to have much bearing on the availability or prices of any of the energy sources described in Chapter 4. Their effects lie more in the direction of affecting demand and acceptability. The civil rights and women's liberation movements increase the demand for energy, whereas the environmental movement shifts the demand toward more highly refined fuels of greater acceptability.

However, as we turn to a consideration of some of the political factors that influence world energy activities, we may anticipate significant effects on fossil fuel prices and availability.

CHAPTER SEVEN

POLITICAL FACTORS

If all oil fields were known, or at least well suspected, as tends to be more or less true for the lower 48 United States, then the major costs associated with bringing oil to market would be development costs for converting known or well-suspected resources to reserves, plus transportation costs. Capital and operating costs plus a minimal return for the marginal production would determine price. As the richest fields began to be depleted, costs would rise as it became more difficult to develop new reserves in the partially depleted fields or as it became necessary to develop less desirable fields. The marginal cost would tend to rise and the price would tend to rise. At the same time, the economies of scale and new technology, working in the opposite direction, would reduce costs, and the price would tend to fall. These opposing factors have been major forces in recent years, with the overall result (in combination with new discoveries as discussed in the next paragraph) that prices have declined gradually when measured in constant dollars.

If it were not for the possibility of new discoveries, this would be the whole story. Producers, knowing what lay in the ground, knowing the economies of scale and the cost-cutting potentialities from new technology, and estimating future demand, could plan ahead rationally for appropriate growth or shrinkage in productive capacity. There might be misjudgements, but there would be no surprises. However, particularly on a worldwide basis, there remains the very real possibility that deposits of low-cost oil, previously unknown and unsuspected, will be discovered and developed, upsetting the or-

63

derly progression of reserve development anticipated by the petroleum industry.

Suppose, as a first example, that a new oil field is discovered and developed to produce low-cost oil with a flow equal to 1 percent of United States oil production. Because the oil is low-cost oil, it will surely be profitable to market and will substitute for marginal production which must decline by 1 percent of United States production. The new marginal production, which was not quite marginal before, is very slightly lower in cost, and prices can drop very slightly. This slight drop in price is hardly noticed by the developers of the new oil, and the only parties discomfited by the new oil are the marginal producers who were squeezed out. Yet their discomfort is short-lived, for with United States oil consumption growing at 7 percent per year, they are back in business again in less than 2 months. More realistically, the marginal producers are subjected to an intermittent barrage of small new discoveries, tempered by a steady growth of demand, so they are able more or less to plan their future. And new discoveries, as long as they continue to be made, contribute an additional force tending to reduce costs and prices.

As a second example, imagine that a much larger low-cost oil field is discovered and developed to produce a flow equal to 20 percent of United States oil production. Again (assuming that other oil flows such as imports and exports remain constant), this oil substitutes for what was previously marginal oil, and 20 percent of the old United States production is shut down. The new marginal producer has noticeably lower costs, and prices drop noticeably, although the developers of the new oil are still doing very well. The squeezed-out marginal producers are not at all happy, because even with demand growth at 7 percent annually, some of them will be out of business for nearly 3 years. This possibility is inherent in the development of Alaskan oil, which fits the example pretty well. For good reasons to be discussed later, however, it is more likely that Alaskan oil, if and when it is allowed to flow, will displace imports rather than displace other United States production.

As a third example, imagine that an extremely large and very low-cost field is discovered and developed to produce a flow equal to 10 times the production of oil in the United States. This is a disaster for existing United States producers, most of whom are driven out of business for decades. It is not all that rosy for the developers of the new field either, because the price drops so far that their revenues are less than if they had held back on development, restricted production, and thus kept prices from falling so far. (From the producer's standpoint, it is better to sell half the volume of oil at twice the price, since costs are less and revenues are the same; and even better to sell half the volume at 3 times the price.) Were a new field of this type to be discovered in the United States, where competition tends to be the rule, particularly if a number of producers shared production rights to the new field, there is little doubt that the price of oil would plummet, some fortunes would be lost and others would be made, and the consumers of petroleum products would benefit. In point of fact a new field of this type has been discovered in the Middle East, with tremendous and nerve-racking potential impacts on world energy activities.

Middle Eastern oil reserves amount to about 10 times United States oil reserves, and to more than half of the entire world's reserves. The incremental supply cost of Middle Eastern oil is about 10¢ per barrel at dockside in the Persian Gulf, including a 20 percent return on all necessary investments. In 15 years, even with no new discoveries and with no new technology, the incremental supply cost is not likely to rise beyond 20¢ per barrel (7.1). There is no doubt at all that costs will remain far below the present Persian Gulf price for decades, even with vastly increased output.

While oil production was building up in the Middle East in the early years, the United States was a major supplier of oil to Europe. The price in Europe reflected the United States Gulf port price of oil plus tanker transport to Europe. This was the price at which Persian Gulf oil also sold in Europe after tanker transport through the Suez Canal (and it is from working backward from the price of oil in Europe, subtracting known tanker and Suez toll costs, that Persian Gulf oil prices have been deduced). No Middle Eastern oil flowed to the United States, because it commanded a higher price in Europe and cost less to transport there.

The situation has continued to change over the years. The flow of oil from the Middle East grew to the point where it supplied most of Europe's demand, although the price in Europe stayed high, in part because of European actions to protect their coal industries. The 1967 Suez crisis closed the canal and spurred the construction of large tankers to carry oil around Africa, and the cost of tanker transport declined so far that Middle Eastern oil gained a substantial ability to undersell domestic oil in the United States, although most of it was kept out by restrictions on imports. The potential now exists for substantial further increases of Middle Eastern production, for a collapse of world oil prices, for the ruin of much of the United States oil business, and for the reimpoverishment of the Middle Eastern countries, as in the third example discussed. Yet except possibly for United States consumers, and probably even including them as is discussed subsequently, all parties recognize the dangers and may be able to avert them.

The United States enacted effective oil import restrictions that kept out a flood of Middle Eastern oil at a time when much money could have been made by bringing it in—under whatever degree of competition or monopoly may have existed among the oil producing companies and countries (7.1). Oil import restrictions allowed the production of domestic oil to continue to grow so that marginal producers were not forced out. They allowed the production of domestic coal to continue at an approximately constant level. Businesses were saved from collapse and jobs were saved from extinction. Europe has followed a similar path in protecting coal. It is likley that the United States will continue to protect domestic fossil fuel industries in the future as it has in the past.

When the Persian Gulf price was $1.25 per barrel, there remained an excess "rent" (as the economists call it) of $1.15 after deducting the 10¢ cost. At first the oil companies that held Middle Eastern concessions kept most of this rent, but gradually over the years the governments of the Middle Eastern countries kept increasing their share, until, after combining forces in the Organization of Petroleum Exporting Countries (OPEC) in 1960, they gained the lion's share. Although OPEC may have been formed to help transfer the oil rent from the oil companies to the producing countries, its motivation can be expected to shift toward maintaining the total oil rent at as high as possible a value.

One of OPEC's goals is to restrict oil production to a degree such that the price will not collapse and the total rent will be maximized. Some economists question their ability to do so, when each member country is faced with the temptation to shave prices, gain market share, and gain a larger share of the rent. Of course they cannot all do it, for a series of such actions would drive the price down to where they all lose. But because they know this OPEC may find a way to limit and allocate production. They may be helped by consuming countries, whose interests are very similar, as we have seen.

With both producing and consuming countries in favor of preserving prices, each group for its own reasons, who then is not in favor? One possibility has been mentioned: the producing country that feels that its production allotment under OPEC has been set too small, and who may try to break away. Another possibility is the consuming country that has no domestic fuel industry to protect, such as Japan. It may be in time that Japan and one or two disgruntled OPEC countries may negotiate a Japanese oil price lower than the world oil price and independent of it. A third possibility is the United States consumer, who will be paying more for oil than he would under world competition. But the alternative of world competition across national boundaries is not realistic. Who is to say what would happen to prices if the OPEC countries were actually, through vigorous competition, to drive most of the domestic United States oil industry into bankruptcy and the nation's oil-producing infrastructure were lost? History suggests that the then-monopolists might not be able to resist their temptation to raise the price as high as the traffic would bear in the knowledge that the competition had been destroyed. And it would be difficult for the oil infrastructure to rise again in the United States, for who would invest money in it knowing that the monopolists could wipe them out again? Overall, the consumer may well be better off to pay more for oil now in exchange for a sure supply at a reasonable price later.

This discussion of the world oil industry has been oversimplified and idealized, but I believe its major elements are correct. The fundamental political factors relating to Middle Eastern oil and the United States are three in number:

1. The desire of the OPEC countries to limit production, keep prices from collapse, and maximize their total rent from the United States (rent per barrel sold to the United States, multiplied by the number of barrels sold to the United States).
2. The desire of the United States to protect its domestic oil and coal industries for a variety of reasons, including preservation of jobs, protection of investments, and preservation of the industrial infrastructure.
3. The potential exposure of United States consumers to monopolistic powers beyond the control of their government.

These three factors relate to the direct two-party interests of the United States as an oil-consuming region and the Middle East as an oil-producing region, and are independent of the interests of other consuming regions. A fourth significant factor is the ability of oil-producing nations to use price increases and production cutbacks as political weapons against friends and allies of the United States, in an attempt to influence United States policy indirectly. Here lies a motivation for the United States not only to restore the capability for self-sufficiency, but to go beyond self-sufficiency and restore limited capability for supplying friends and allies.

CHAPTER EIGHT

AVAILABILITY OF FUELS

This chapter attempts a synthesis of the geologic, economic, social, and political facts and forces impinging on the energy activities of the world in order to estimate the future availability and price of fossil and nuclear fuels. First, I want to isolate and set aside the United States energy crunch of the early 1970's, a transient phenomenon with limited long-range significance, and then address more fundamental long-range problems.

A large part of the United States energy crunch of the 1970's has been the result of unanticipated social factors, in particular the environmental movement, as has been discussed in previous chapters. The demand for distillate oil turned out to be much higher than anticipated as coal-burning power plants shifted to oil, as electric utilities substituted oil-burning gas turbines for delayed nuclear reactors, and as automobiles began burning more gasoline because of newly legislated safety and antipollution equipment.

Natural-gas price regulation has been another contributing factor. Undertaken to help keep the price of gas low for residential consumers, regulation of the wellhead price of natural gas intended for interstate shipment interfered gradually but progressively with competitive interfuel substitution and pricing. In 1970 the owner of a gas well got only about a third of the money from selling gas as did the owner of an oil well from selling oil with the same heating value. The motivation for developing new gas reserves eroded, and the supply began to decline. At the same time the increasing desirability of gas, relative to other fuels, caused demand to soar. With supply and demand decoupled from their customary meeting ground in the marketplace, they moved so far out of balance that rationing, imports, and synthetic gas have become necessary. But it appears that the underlying cause of much of the imbalance—the artificially low price set for interstate natural gas—is now more widely understood and appreciated, and the price of natural gas will be allowed gradually to seek its competitive level. With the price and market mechanism back at work, we can expect supply to increase and demand to decrease until, after several years of adjustment, the two again come into balance.

Uncertainty about the political factors surrounding Middle Eastern oil may have contributed to the energy crunch to a lesser, indirect extent. The potential flood of low-cost oil imports raised the possibility that Middle Eastern oil might underprice much United States production and make it uneconomic. It would be understandable if United States producers were to postpone new investment in developing their high-cost reserves and trim inventories until the import policy was clarified, when they could more nearly count on prices not collapsing. It would be understandable if investment in new oil refinery capacity and in development of coal refining were postponed for the same reason. Now that oil import policy is more clearly defined, such inhibitions on investment should be lessened.

The energy crunch of the 1970's was caused by factors that are so clearly identifiable, so clearly understood, and so readily compensatable through the ordinary response mechanisms of government, industry, and technology that there can be little doubt of its transitory nature. There are real shortages of distillate fuel now that the crunch is on, and there are sharp price rises associated with the shortages, but I expect that the crunch will be over in 2 or 3 years and that prices will drop back to lower levels. With this comment I would like to set aside the energy crunch and consider some of the longer-range effects of the environmental movement on fuel availability.

One long-range effect is the effect on prices. In saying that prices will drop back to lower levels after the energy crunch is over, I do not mean to imply that all prices will drop back to where they were before. The environmental movement is likely to have permanent effects on some price levels, particularly the price of coal. In the old days coal had monetary value more or less in proportion to its heating value. Sulfur burns hot and was not entirely unwelcome as a constituent of coal. Now with restrictions on the emission of sulfur into the air, and with the high cost of sulfur removal from coal or stack gases, coals with high sulfur content are less valuable than they used to be. The price level of high-sulfur coal is sure to adjust appropriately downward. On the other hand, since natural gas and most refined petroleum products contain little or no sulfur, we may expect little adjustment in their price levels on this account.

From the long-term standpoint, a major result of the environmental movement thus appears to have been to diminish the value of high-sulfur coal, to reduce its price while leaving the cost of mining it unchanged, and in all probability to close down a number of marginal Appalachian high-sulfur coal mines. Some of the former and prospective burners of high-sulfur coal will switch to low-sulfur oil, and others will switch to low-sulfur coal from the Western states. Environmental pressure against coal may well continue, particularly against some forms of strip mining, so that the cost of mining coal will increase. However, I anticipate that the major force of the environmental movement will be spent when high-quality mine reclamation becomes mandatory. Certainly in West Germany—where open pit coal mines are currently being reclaimed, even reconstituted, to better farmland and better villages than were there before—experience shows that adequate reclamation is possible and can meet with general public acceptance.

Hence, in the synthesis of impinging factors affecting energy activities, it seems likely that the environmental movement will leave two lasting commandments:

Only refined fuel shall be burned, and
Lands disturbed by mining shall be reclaimed.

Strictly speaking, the commandment on fuels will require that the level of harmful combustion products allowed into the air shall be limited, which allows the possibility of refining the fuel after it is burned rather than before. This approach may be taken by some industrial consumers, particularly in the short run, but it is likely that in the long run it will prove cheaper to refine the fuel first. It is too bad in one sense that both these blows fall on coal, already reeling under the competition from fluid hydrocarbons, but in another sense the coal industry can look back on these final blows with a sense of relief that the worst has happened. The long retreat may be over. Henceforth the economies of scale and new technology, particularly the technologies of coal refining and land reclamation, backed by reserves and resources that will long outlast oil and gas, promise lower costs and growing markets.

Let us turn now to the influence of political factors on the availability and price of fuels. These factors relate primarily to Middle Eastern oil, because of its tremendous abundance and very low cost. Past United States energy policy has been strongly protective of domestic fuel production, as discussed in Chapter 7. The motivations have been to save jobs and to protect investments.

Neither of these motivations will fade away in future years, and they are being augmented by the desire to preserve a domestic petroleum infrastructure and to protect future consumer interests. Hence I feel confident that future oil imports will be restricted.

At one extreme the United States could erect an impenetrable barrier against importing oil or any other fossil fuel, and could go its own way in complete self-sufficiency. This would be a somewhat higher-cost route than if some imports were allowed, but it is entirely possible and feasible. The transition to total self-sufficiency from today's partial dependence on imports could be accomplished within a decade. (This possibility, combined with preparations for carrying it out, might become a useful element in future negotiations with oil-producing nations.) At this extreme the price of fuel in the United States would be just the cost for the existing marginal United States producer, including a minimal return.

At the other extreme the United States could import oil or other fossil fuels to make up all the country's growth in energy consumption. Domestic fossil fuel production would remain unchanged, and a minimum acceptable protection of jobs, investments, and infrastructure would be achieved. Domestic interfuel competition would still be possible, so that coal might make a partial comeback against gas and oil. At this extreme (as for the other) the price of fuel in the United States would be just the cost for the existing marginal producer including a minimal return.

For either extreme, or for any intermediate level of imports, we see that United States fossil fuel prices would be determined by the costs of existing marginal United States producers. This is a consequence of the assumed protection of fossil fuel producers through the imposition of some level of import restrictions. Because the producers are protected, they survive, and because they survive, their marginal costs determine prices. These costs and prices are analyzed in Chapter 5, including experience curves showing how various cost components have declined or might be expected to decline with cumulative experience. If United States fossil fuel production were to shrink, as would be possible with massive imports, these experience curves might be poor guidelines for estimating future prices, because high-cost producers and high-cost production would be driven out of business and the new marginal producers among those remaining in business would have much lower costs. But if imports were restricted so that domestic fossil fuel production were not to shrink, but were at worst to stay constant and at best to continue to grow, as in the two extremes of industry protection just described, then the experience curves in Chapter 5 might be reasonable guidelines for estimating future prices.

We saw in Figure 5.1 how gasoline processing costs have declined with cumulative production. I expect that they will continue to decline along the same experience curve trajectory, and that the costs of processing other refined products will decline similarly. We saw in Figures 5.3 and 5.4 how the cost of crude oil can be anticipated to decline with the advent of a synthetic crude oil industry, and I anticipate that a synthetic crude oil industry will take off at whatever time in the future the price of natural crude oil begins to rise, or perhaps even sooner. This makes plausible a continued decline in the price of fluid hydrocarbons. Similar reasoning makes plausible a continued decline in the price of nuclear fuels. The projected prices depend on the additional cumulative production, and all price trends are downward (in constant dollars) with additional production. This is a different type of price decline than would be caused by massive imports. Imports drive down domestic prices simply by displacing high-cost marginal production, and prices rebound whenever imports cease. The price decline associated with the accumulation of experience comes about more gradually and more permanently because of a decline in the cost of marginal production itself. Since production would accumulate more rapidly if imports were prohibited, marginal costs and prices (particularly of synthetic oil) might drop more rapidly if imports were prohibited.

We have already considered gas turbine production, which was vastly speeded by the need to fill the generation gap caused by nuclear delays, with the result that experience accumulated much more rapidly and costs and prices declined much more rapidly than they would have without the speedup. Substantial cumulative production was put behind the gas turbine industry, and the resulting lower costs and prices are here to stay. Gas turbines became stronger competitors to other forms of generation. The situation with respect to fossil fuels is much the same. Because of import restrictions, price within the United States is determined by the costs of the marginal United States producer. The more we produce, the more experience is accumulated and the lower the cost of recovering natural fuels or manufacturing synthetic fuels. The sooner we achieve a given cumulative production, the sooner we achieve the corresponding lower price, as long as cost and price decline with cumulative production. This process is slowed by imports; thus the price decline is slowed by imports.

If the price of fuel might indeed be less without imports, then why allow imports at all? There are at least two reasons. First, the faster we use domestic fossil fuels, the sooner they are gone. The price drops faster because experience accumulates faster, but it will turn around and rise again sooner because the resources will run out sooner. We have seen that this is a matter of 500 years or so at the earliest for fossil fuels, and much longer for nuclear fuels. Although not a problem of immediate concern, it is a factor to be considered. Second, there can be additional financial impact on consumers of fuel beyond the price they pay for it. For example, a tariff on imports would provide a revenue stream for the Federal government, and other things being equal this would reduce consumers' Federal taxes by the same amount. Insofar as consumers wish to reduce their aggregate outlay for fuel and taxes, they may wish the country to import fuel subject to an appropriate tariff.

A tariff on imported oil is now a reality (although it is called a fee), and I anticipate that it will continue. Yet a tariff alone may not be adequate to limit imports, for any producing country that wanted to expand production and increase sales to the United States could simply lower its price so that price FOB the Persian Gulf, plus tanker charges, plus tariff, still amounted to less than the United States price, and could in this way still manage to drive much of the United States petroleum industry out of business. Stability requires either a very high tariff—so high that only a portion of Middle Eastern oil would be profitable at the very low price it could command after payments for transport and tariff—or it requires quotas. Import quotas can be allotted to various importing agencies (such as oil companies), which gives the Federal government some power over the importing agencies and the importing agencies in turn some power over the exporting countries. Or export quotas can be allotted to exporting countries, which gives the Federal government some direct power over the exporting countries and removes certain temptations inherent in the other alternative. Overall I am inclined to pick the second alternative—allocation of a quota to each exporting country—as the one to which the United States will turn in the long run. Yet either way the combination of quotas and tariff achieves stability of imports. With a high enough tariff it saves the country money overall and stretches domestic resources. The ability to adjust both quotas and tariffs enables the balance of payments to be kept under control and gives the United States bargaining power over the petroleum exporting countries.

The idea that fossil fuel prices, measured in constant dollars, will continue to decline for a very long time (with temporary ups and downs such as the up of the current energy crunch being followed shortly by a compensating down), although based on evidence and analysis that persuades me, may not appeal to all readers. Let us therefore consider the alternative possibility, namely, that prices will rise as production accumulates.

There is no effect on the analysis supporting the concept that a combination of tariffs and quotas will lead to stability of fossil fuel imports, reduction of balance of payments problems, and reduction of taxes on United States consumers, and will give the United States bargaining power over petroleum-exporting countries. The United States can save money overall by importing oil, and can at the same time have the upper hand in negotiating for it. The major differences are that imports cause prices to rise less rapidly (instead of causing them to fall less rapidly) and the fallback bargaining position of completely cutting off imports is more expensive. The basic fact remains that the nation is blessed with an abundance of fossil fuel resources. With several years lead time for the necessary incremental construction we can return to self-sufficiency at any time. Hence the price of oil in the United States cannot exceed the equilibrium price for self-sufficiency any longer than it takes to develop the required additional reserves and productive capacity.

To summarize not only this chapter but much of the book to this point, there is every reason to believe that domestic fossil fuels will continue to be available, at slowly declining prices (in constant dollars), in adequate amounts to support total anticipated energy consumption for many centuries. And there is every reason to believe that domestic nuclear fuels will continue to be available, at slowly declining prices (in constant dollars), in adequate amounts to support total anticipated energy consumption for many millenia. Insofar as nuclear fuels or other energy sources take an increased proportion of the load, fossil fuels will last correspondingly longer.

CHAPTER NINE

ROLE OF ELECTRICITY

Electricity is not a primary source of energy, but rather the most highly refined form of energy. The energy content of coal, oil, gas, falling water, or uranium can be transformed to electricity, and the history of energy consumption over the past century shows a steady growth in the proportion of these primary energy sources that have been so transformed prior to final use by consumers. There is no alternative to electricity for some purely electrical and electronic end uses of energy. But for most other end uses of energy consumers have a choice of burning fuel on their own premises or of utilizing electricity generated from fuel burned off the premises in an electric power plant. Consumers have opted increasingly for electricity.

The major end uses for electricity in the United States, ranked in order of the electric energy they utilized in 1968, are given in Table 9.1 (1.2). The category "Other" includes small appliances, computers, elevators, escalators, office machinery, commercial electric heat, industrial air conditioning, and other miscellaneous end uses.

Some end uses such as industrial drive are almost completely electrified, whereas others such as transportation are almost completely unelectrified. In viewing the pattern of electrification of the United States, it is instructive to classify end uses into groupings characterized by the degree of electrification, as is done in Table 9.2. Three groupings emerge: a group of completely electrified end uses that together account for 22 percent of energy consumption but

TABLE 9.1   END USES FOR ELECTRICITY
(UNITED STATES, 1968)

End Use	Percent of Electricity
Industrial drive	39.7
Refrigeration	11.6
Lighting	10.8
Water heating	6.2
Electrolytic processes	5.8
Air conditioning	5.6
Space heating	3.3
Direct heat	2.6
Television	2.6
Cooking	2.1
Clothes drying	1.0
Transportation	0.4
Other	8.3
	100.0

a very large 84 percent of electric energy consumption; a group of partially electrified end uses, listed in order of the degree of electrification, that account for 38 percent of energy consumption but only 15 percent of electric energy consumption; and a group of unelectrified end uses that account for 40 percent of energy consumption but essentially no electric energy consumption. By considering the characteristics of these various groupings, we can gain some insight into the process of electrification.

First consider the electrified end uses, for which the superiority of electricity over direct combustion of fuel on the premises is so great that direct combustion on the premises has virtually disappeared. Some of these end uses are uniquely electrical and would not otherwise be possible: electrolytic processes, television, and some of the "Other" category such as telecommunications and modern computer processing. In aggregate these uniquely electrical uses consume perhaps 3 percent of the energy and 11 percent of the electricity in the United States. Another group shares in common the utilization of electric motors to perform work tasks of various sorts. Here a choice can be made— steam engines, water wheels, or diesel engines on the premises; or electricity on the premises with the fuel conversion in a distant power plant—and electricity has been chosen. The group of end uses includes industrial drive, refrigeration and air conditioning (both of which are powered by motors), and some of the "Other" category, for a total of about 15 percent of the energy and about 62 percent of the electricity consumed. Illumination is the third type of electrified end use, consuming about 3 percent of the energy and 11 percent of the electricity. The choice of electric illumination over direct combustion of tallow, oil, and gas is nearly universal.

In summary, the fully electrified end uses can be classified into three groups: those that are possible only through the agency of electricity (11 percent of electricity consumption), those that involve stationary motors or engines performing work of some sort (62 percent of electricity consumption), and illumination (11 percent of electricity consumption). The electronic group is electrified because there is no alternative. Illumination and the stationary

TABLE 9.2   ELECTRIFICATION OF THE UNITED STATES (1968)

End Uses	Percent Electric	Percent of United States Energy	Percent of United States Electricity
Electrified			
Industrial drive	~100	10.3	39.7
Refrigeration	~100	3.0	11.6
Lighting	~100	2.8	10.8
Electrolytic processes	100	1.5	5.8
Air conditioning	96	1.5	5.6
Television	100	0.7	2.6
Other	~100	2.2	8.3
	~100	22.0	84.4
Partially electrified			
Clothes drying	70	0.4	1.0
Cooking	40	1.4	2.1
Water heating	38	4.2	6.2
Direct heat	6	11.0	2.6
Space heating	5	20.8	3.3
	11	37.8	15.2
Unelectrified			
Transportation	—	25.6	0.4
Process steam	—	14.6	—
	—	40.2	0.4

~ indicates approximate percentage.

engines have been electrified because the performance is better, the cost lower, the convenience greater, fuel is conserved, and pollution is lessened, as is discussed in detail in the next chapter.

Next consider the unelectrified end uses, process steam and transportation. There is no motivation for electrifying process steam, for the usual way of making electricity is to make steam first and then transform part of the energy of the steam to electricity with the rest being lost in the process. The sensible thing is to utilize the steam for industrial processes the first time it is made, and not to change it to electricity at extra cost and energy loss only to change it back. Transportation resists electrification because the engines are not stationary, and it is difficult to devise an arrangement of wires to supply electricity to a moving motor. If the vehicle runs on rails or some other well-defined track, electric wires can be run parallel to the track and electrification is possible. But the trend in the United States has been in the opposite direction for half a century, with vehicles of all sorts—automobiles, trucks, aircraft, ships—all increasingly free to go their own ways. Electric storage batteries carried on a vehicle offer one possibility for electrification of transportation, but aside from low-speed, short-haul uses such as forklift trucks and golf carts, this method has not made much headway. For highway transportation, the development of the internal combustion engine has pulled ahead of battery development, so that electric automobiles have been losing ground since the 1890's.

Finally let us consider the partially electrified end uses, whose common feature is the generation of heat. Here the penetration of electricity is small, particularly for the largest end uses. The reason for the limited penetration of electricity into these heating applications lies in the relative inefficiency of some forms of electric heat. Where convenience is particularly important and energy consumption is modest, as in clothes drying, cooking, and water heating, electricity has won moderate acceptance. But where consumption is greater and cost looms larger relative to convenience, direct combustion of fuel on the premises is ordinarily the rule.

The efficiency with which electric energy is generated and utilized in its various end uses is of such great importance to understanding the past and probable future course of electrification that it merits a careful analysis. The next chapter, directed to this purpose, is somewhat more technical than the rest of the book, and nontechnical readers may feel like skipping it. But I encourage them not to. They may miss a few details, but I hope they may sense the overall picture.

CHAPTER TEN

EFFICIENCY OF ENERGY UTILIZATION

As we have seen, the major uses for energy are for the production of work or heat, and it is important to understand the relationships between them. Both work and heat mediate the flow of energy from one place to another. (Other agencies also can mediate the flow of energy, but for now we consider only work and heat.) The flow of energy is called work when it exerts a force (for example, the force exerted on a golf ball by a swinging golf club while the two are briefly in contact, energy thereby flowing from the club to the ball). The flow of energy is called heat when it does not exert a force (for example, the flow of energy from a hot oven to a cold potato). Work tends to change the state of motion of the object to which it flows, and it is useful for pushing things around or turning wheels. Heat tends to change the temperature of the object to which it flows (as in heating water) or to change its nature (as in melting iron), and it is useful for such purposes.

Because work and heat are alternate modes for the flow of energy, they can be quantified by the amount of energy that flows via each mode. Consider as an example the operation of a modern steam turbine used to drive an electric generator. The fuel burned in the boiler generates 10 billion Btu of energy each hour of operation. A portion of this energy, amounting to 4 billion Btu per hour, flows through the rotating turbine shaft in the form of work, where it is used to turn the shaft of an electric generator. The rest of the energy, amounting to 6 billion Btu per hour, is discharged into the air or into cooling water. This situation is diagrammed in Figure 10.1.

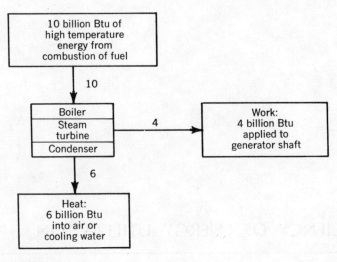

Figure 10.1.

Since the boiler/steam turbine/condenser heat engine did only 4 billion Btu of work when supplied with 10 billion Btu of combustion energy, only 40 percent of the energy in the fuel was actually used for the purpose for which it was intended—namely, to turn the generator shaft. The other 60 percent was lost as unutilized waste heat. The ratio of useful work output to energy input for an engine is called the efficiency of the engine, and in this example the engine has an efficiency of 40 percent.

Many engines, including jet engines, diesel engines, and steam turbines, re-ceive energy at high temperature, transform some into work, and discharge the rest of it as heat at a lower temperature. Much study has been devoted to the potential efficiency of these engines, and the concept of a "reversible heat engine" has emerged as an idealization against which the lesser performance of real engines can be measured. Imagine a reservoir of energy at a high temperature denoted by T_1 and a second reservoir of energy at a low temperature denoted by T_2. Imagine a reversible heat engine to be in contact with each of them and to be able to exchange heat with each of them, and let the engine be provided with a rotating shaft along which work can flow, as sketched in Figure 10.2.

When utilized to generate work, a reversible heat engine draws heat from the high-temperature reservoir, delivers work along the rotating shaft, and dis-charges heat to the low-temperature reservoir. It is the idealization of a steam turbine power plant. Figure 10.3 shows what happens to a quantity Q_1 of heat that flows into the engine from the hotter reservoir: Q_2 of it goes to the colder reservoir and $Q_1 - Q_2$ of it comes out as work.

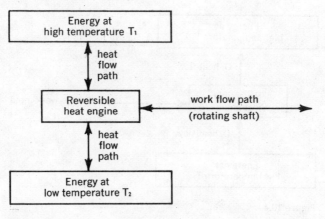

Figure 10.2.

Now suppose that the shaft connected to the engine is made to turn in the opposite direction by means of some outside agency, so that work flows into the engine instead of out. The engine then draws heat from the low-temperature reservoir and delivers heat to the high-temperature reservoir. It is a heat pump, the idealization of an air conditioner or refrigerator. Figure 10.4 shows the flows of work and heat associated with the delivery of a quantity Q_1 of energy to the hotter reservoir: Q_2 of it comes from the colder reservoir and Q_1-Q_2 of it comes in along the shaft as work. This picture is the same as the previous one except that everything is flowing in the opposite direction. This is the meaning of reversibility. (Most real engines do not run backward very well, because they are designed to give their best performance in one direction only. The optimum design for a turbocompressor is some-

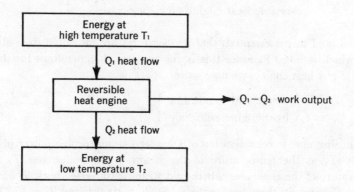

Figure 10.3.

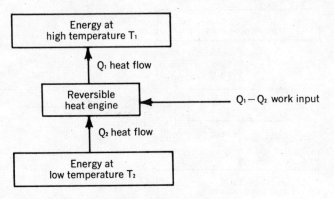

Figure 10.4.

what different from the optimum design for a turbine, for example, although each can be made to run backward badly and do the other's job poorly.)

The science of thermodynamics has ascertained some remarkable properties of reversible heat engines. Their performance does not depend on their internal construction at all, but just on the two temperatures T_1 and T_2. When used to perform work, the efficiency of a reversible heat engine is the maximum possible efficiency achievable by any heat engine, reversible or not, operating between the same two temperatures. When a reversible heat engine is used as a heat pump, the heat delivered to the hotter reservoir and the heat removed from the colder reservoir are the maximum possible amounts, in relation to the work supplied, that can be transferred by any heat pump, reversible or not. The efficiency of a reversible heat engine depends on the temperatures T_1 and T_2 through the relationship

$$\text{reversible heat engine efficiency} = \frac{T_1 - T_2}{T_1}$$

provided (and this is essential) that temperatures are measured from absolute zero, which is $-460°F$. Since this is the theoretical upper limit for the efficiency of any heat engine, we may write

$$\left\{ \begin{array}{l} \text{theoretical maximum} \\ \text{heat engine efficiency} \end{array} \right\} = \frac{T_1 - T_2}{T_1}.$$

Returning now to consideration of a modern steam turbine power plant we identify T_1 as the temperature of the steam in the boiler and T_2 as the temperature of the condenser where cold steam is converted back to water. In degrees Fahrenheit, these temperatures are $T_1 = 1000°F$ and $T_2 = 120°F$ for a representative water-cooled fossil steam plant. The corresponding absolute

(Rankine) temperatures are $T_1 = 1460°R$ and $T_2 = 580°R$. The corresponding theoretical maximum efficiency is

$$\left\{ \begin{array}{l} \text{theoretical maximum} \\ \text{steam plant efficiency} \\ T_1 = 1460°R, \quad T_2 = 580°R \end{array} \right\} = \frac{T_1 - T_2}{T_1} = \frac{880}{1460} = 0.60$$

or about 60 percent. Modern boiler-turbine-condenser practice achieves close to 40 percent thermal efficiency, or two-thirds of the theoretical limit for 1000°F steam. This is a remarkable technological achievement, but engineers still are hard at work trying to improve on it.

Now, having discussed heat, work, heat engines, and efficiencies, I would like to return to the subject of electrification. Much is said in print about the inefficiency of electric power generation—that most of the energy in the fuel consumed in a power plant is lost as waste heat while only a portion is utilized as electric energy. This is all true. But there is an implication that the waste is due to electrification, and that is not true. Every heat engine generates waste heat. Waste heat is an inevitable consequence of the use of engines for the production of work, and it is fairer to tag engines in general, rather than electric power in particular, with responsibility for generating it.

The flow of electric energy is equivalent to the flow of work. Work can be transmitted from one place to another by means of a long rotating shaft, or by means of a belt stretched over pulleys, or much more conveniently, flexibly, and inexpensively by means of electricity along conducting wires. Once generated, electricity can be utilized with very little additional waste.

Consider two alternatives: (1) burning oil in a power plant to make electricity, then transmitting the electricity to a factory where it turns an electric motor that turns a lathe, and (2) burning oil in a diesel engine that turns the lathe directly. The two alternatives are compared in Figure 10.5. The comparison does not use the best power plant with 40 percent efficiency, but an older power plant with 30 percent efficiency more characteristic of the average around the country. It does not use the most efficient, large, modern diesel engine, but an older, smaller diesel engine with 25 percent efficiency more characteristic of the average around the country. Each alternative starts with 100 units of oil combustion energy and traces the subsequent energy flow. From the standpoint of generating waste heat, there is an approximate standoff. The power plant is more efficient than the diesel engine, but there are additional losses in transmission and in the electric motor that tend to even things up. Hence electrification of industrial drive does not increase the amount of waste heat associated with powering industry, but merely shifts its location from factories to power plants.

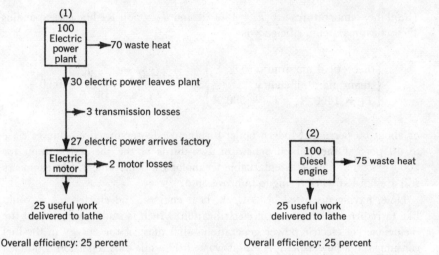

Figure 10.5. Overall efficiencies of electric power and diesel engine power (illustrative). Each alternative starts with 100 units of oil combustion energy and traces the subsequent energy flow.

From the standpoint of overall cost, the comparison favors electricity. Fuel is cheaper for the power plant, and the economy of scale is working for the power plant. The cost of a power plant plus transmission system plus electric motor (per unit of work delivered to a lathe) is less than the cost of a diesel engine (per unit of work delivered to a lathe), partly because of the economy of scale but also in large part because the power plant is more fully utilized. Whereas the diesel engine might run 40 hours a week at an average 30 percent of its maximum rating for an overall utilization factor of about 7 percent, the power plant by providing electricity to a number of users whose demands occur at different and partly overlapping times might have an overall utilization factor of 65 percent, thereby achieving a much better utilization of the invested capital. Operating and maintenance costs are less for the power plant for much the same reason. Factory layout can be rearranged much more easily and cheaply when electricity provides the power. And of increasing importance as the country turns its concern to pollution abatement, large power plants generally are able to burn fuel more thoroughly than the alternative of many small engines, substantially lessening pollution. From this example, it is easy to see how the performance of stationary work—work in factories, refrigeration, and air conditioning—has been nearly completely electrified.

Illumination provides another instance in which substantial benefits and savings have been achieved through electrification. Compared with the alternative of providing illumination by direct combustion, electricity gives more light at less cost with less bother, less pollution, and greater safety. And fuel is conserved, for a gallon of oil burned in a power plant to make electricity that operates an electric lamp gives more light than a gallon burned directly in an oil lamp. Much of the superiority of electric illumination results from the ability of electricity to provide heat at very high temperatures, as discussed in the following paragraphs.

Just as the work flowing along a rotating shaft can be converted to heat through mechanical friction, so too the electrical work flowing along a conductor can be converted to heat through electrical friction or resistance. Electricity can be forced to pass through a section of poor conductor where high electrical friction or resistance concentrates the generation of heat, and such electric resistance heaters have many applications. Resistance heaters take many forms, including the ribbon of resistive metal alloy in a toaster, the gap of resistive plasma (partially ionized gas) in the carbon arc used to melt iron in an electric arc furnace, the plasma in a mercury vapor lamp, and the tungsten filament in an incandescent lamp.

When fuel is burned in air, the air temperature rises as combustion energy is released, until finally the oxygen is used up, the release of combustion energy ceases, and the temperature stops rising. The ultimate combustion temperature that can be reached differs for different fuels, but is in the neighborhood of 3500°F for the fossil fuels. If combustion is carried out in pure oxygen instead of air, so that no energy is wasted heating inert atmospheric nitrogen, the temperature can be raised to about 5000°F. Special fuels such as cyanogen can be used for somewhat higher temperatures, but the end of the line is soon reached. Whenever still higher temperatures are required, we must turn to other means than combustion. Electric resistance heating is of particular value at these very high temperatures, since it can deliver heat at any temperature, no matter how high, including the 10,000°F plasma in a mercury vapor lamp and even the hundred-million-degree plasmas being studied in thermonuclear fusion research. Resistance heating stays competitive down to lower temperatures such as the 2800°F at which iron melts, even though these temperatures can be reached by combustion, because combustion heating tends to lose efficiency through hot combustion products going up the chimney.

At low-to-moderate temperatures, however, the efficiency of combustion heating tends to improve because the products of combustion can give up a larger proportion of their energy as they cool down to a lower temperature before going up the chimney. For end uses such as space heating, water heating, cooking, and much industrial direct heat, direct combustion is an inexpensive and efficient means for providing heat. Electricity has found only limited application to these end uses, and it is instructive to compare the two methods of electric heating—resistance heating and the electrically driven heat pump—with direct combustion to see why electricity has made so little headway.

When we discuss the performance of a heat pump, we are interested in the ratio of the heat delivered to the hot reservoir (the desired result) to the work required to operate the pump (the necessary input). This ratio is called the coefficient of performance (COP) of a heat pump. Symbolically,

$$\text{COP of a heat pump} = \frac{\text{heat delivered to hot reservoir}}{\text{necessary work input}}$$

When a reversible heat engine is used as a heat pump, its COP is simply the reciprocal of its efficiency when used as an engine, the latter being, of course, the ratio of the work output to the heat received from the hot reservoir. Hence

$$\text{reversible heat pump COP} = \frac{T_1}{T_1 - T_2}.$$

This relationship shows the theoretical maximum amount of heat that can be pumped into a reservoir at temperature T_1 per unit work input.

As an example, consider the possibility of heating a house by means of a heat pump. Suppose that the outside air temperature is 20°F = 480°R and the inside temperature is 70°F = 530°R. Then the maximum theoretical ratio of heat (delivered into the house) to work (required for doing it) would be

$$\left\{ \begin{array}{l} \text{theoretical maximum} \\ \text{heat pump COP} \\ T_1 = 530°\text{R}, \quad T_2 = 480°\text{R} \end{array} \right\} = \frac{T_1}{T_1 - T_2} = \frac{530}{50} = 10.6.$$

The work put into running the pump would be multiplied more than tenfold in delivering heat to the house. For each 10.6 units of heat delivered indoors, 1.0 unit would come from the work expended in driving the heat pump and 9.6 units would come from the air outdoors. However, this is the theoretical limit, and just as heat engines designed for producing work do not come up to the theoretical efficiency, so too heat engines designed for pumping heat do not come up to the theoretical performance. At the level of development so far achieved, a commercial heat pump working between 20°F and 70°F is able to deliver only about 3 times as much energy in the form of heat as the energy content of the electricity that drives it. This is far short of the theoretical multiple of 10.6, but it is still of significance for space heating.

With today's technology, a comparison of space heating by direct combustion, electrically driven heat pump, and electric resistance heating looks ap-

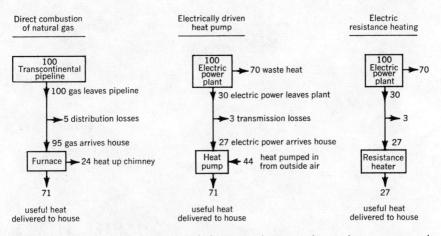

Figure 10.6. Overall performance of direct combustion, electric heat pump, and electric resistance heating (illustrative). Each alternative starts with 100 units of natural gas energy and traces the subsequent energy flow.

proximately as shown in Figure 10.6. Each alternative starts with 100 units of energy in the form of bulk fuel and traces the subsequent energy flow.

We see in this comparison some of the reasons why low-temperature heating has not been electrified to any great extent, and we see some of the potential for increasing electrification in the future. From the standpoint of fuel conservation and overall cost, electric resistance heating is not attractive. More fuel is required, and the cost is greater. Yet where only small amounts of heat are desired and where convenience is an important consideration, resistance heating is frequently chosen. Heat pumps and direct combustion stand more or less at a draw when fuel conservation is considered, but heat pumps are more expensive than furnances. In centrally air-conditioned buildings this does not matter as much, because a heat pump can both cool and heat, but it remains a factor in favor of direct combustion.

For electric heating to make good solid inroads against direct combustion, two lines of technological progress are essential. The efficiency of electric generation from fossil fuels must be improved, and the performance of heat pumps must be improved. Improved efficiency of generation would move the trade-off between electricity and combustion for industrial heat to lower temperatures, making electricity attractive for more industrial heating applications. And it would increasingly tilt the decision toward electricity for applications such as cooking and water heating, where convenience is an important consideration. Improved heat pump performance would lead progressively to space conditioning that is more conservative of fuel, less expensive, and more convenient in comparison with direct combustion.

CHAPTER ELEVEN

PROGRESS OF ELECTRIFICATION

The past century has seen a steady worldwide growth in the consumption of electricity. When measured in terms of kilowatt-hours of electric energy (1 kilowatt-hour equals 3412 Btu) as is the practice of the electric utility industry, United States and world consumption have been growing at an average rate of about 7 percent for half a century, as shown in Figure 11.1. This growth rate corresponds to a doubling of electric energy consumption every decade.

It is important to keep in mind that most of the growth in electric energy consumption has resulted from shifts away from other forms of energy consumption. Waterpower used to be harnessed directly to factory machinery by waterwheels, pulleys, and belts, but now it is harnessed indirectly through electricity. Hydroelectric power has substituted for direct waterpower. Fuels used to be burned on the users' premises for illumination, stationary work, and heat, but increasingly fuels are burned off the users' premises in electric power plants. Electric power is substituting for direct combustion. As electric energy is increasingly chosen over direct fuel combustion, a growing proportion of energy is converted to electricity prior to consumption. By 1970 this proportion had risen to nearly 27 percent in the United States as shown in Figure 11.2. Since the proportion of energy turned to electricity cannot increase beyond 100 percent and may indeed level off well short of 100 percent, there is a prospect that the high growth rate of electricity may come to an end within a few more decades.

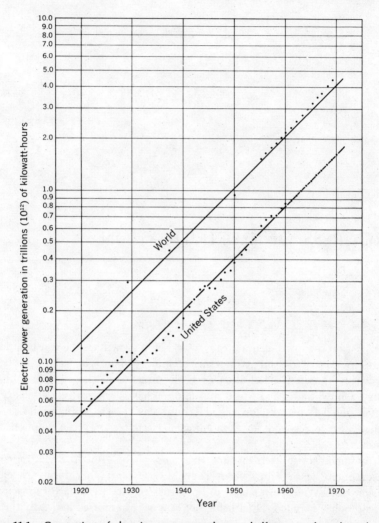

Figure 11.1. Generation of electric power over the past half century, for selected years worldwide (11.1) and annually for the United States (11.2). The trend lines correspond to uniform growth at 7.2 percent per year (100 percent per decade). World electric power generation has been growing at about 8 percent per year over the past decade.

The efficiency of electric generation increased over the years as boiler, turbine, and generator technologies were continually improved, as shown in Figure 11.3 (the pause during the 1960's will be discussed in the next paragraph). This improvement in efficiency was one of the factors that progressively increased the attractiveness of electricity. As the cost of electricity

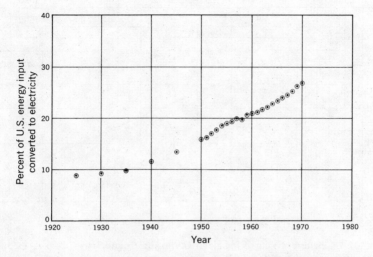

Figure 11.2. Conversion of primary energy into electric energy, the United States 1925 to 1970. The percentage of primary energy input converted to electricity grew steadily from about 9 percent in 1925 to nearly 27 percent in 1970 (11.3).

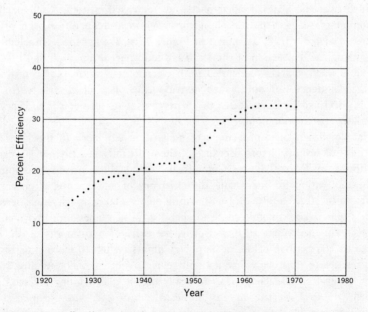

Figure 11.3. Overall efficiency of converting fossil fuel energy to electric energy achieved by the United States electric utility industry for the years 1925 to 1970 (11.2). The newest and most efficient power plants have efficiencies close to 40 percent, but the system average reflects the performance of a mix of power plants of different ages.

95

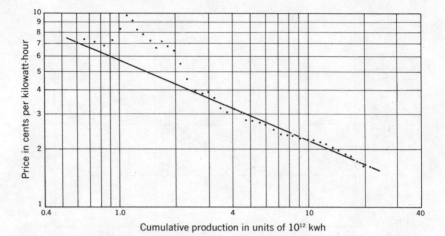

Figure 11.4. Experience curve for electricity in the United States, 1926 to 1970 (11.2). The trend line corresponds to a 25 percent decline in price for each doubling of cumulative production. Price rose above the trend line during the great depression.

declined over the years, helped partly by increased efficiency and partly by economies of scale in power plant construction and operation, the price of electricity also declined, as shown in Figure 11.4. Except for a transient price rise during the depression of the 1930's, due in part to the high and relatively inflexible fixed charges associated with electric generation that continued in the face of depression-diminished demand, the price of electricity (in 1970 dollars) declined by 25 percent, to 75 percent of its former value, every time cumulative production doubled.

The slowdown in improvement of generation efficiency in the 1960's was the result of several factors. Because of the artificially low price of natural gas resulting from Federal Power Commission wellhead gas price regulation, part of the motivation for increasing the efficiency of gas-burning electric power plants (with 30 percent of all fuel-burning power plant capacity) was removed. It does not make economic sense to spend a lot of money on a more efficient power plant to save cheap fuel. But as the price of natural gas is allowed to rise to its competitive value, or as power plants switch to more expensive oil, the pressure to improve efficiency will revive. Another factor tending to hold down generation efficiency was the heavy use of gas turbines necessitated by nuclear delays. As the magnitude of the nuclear and pumped-storage generation gap became apparent, more and more low-efficiency gas turbines were rushed into service. Ideally, for optimum overall system performance and lowest electric power cost, gas turbines would be operated only a few hundred hours per year at times of peak demand for electricity, so that their low efficiency would hardly affect overall system efficiency (while their low capital cost would more than compensatingly reduce the overall cost of electricity, as is discussed more fully in Chapter 13). But in the emergency situation of the 1960's, gas turbines had to be run for thousands of hours each year, significantly diminishing overall system efficiency. When nuclear and pumped-storage plants are placed in commercial operation and the emergency is over, system efficiency will recover as gas turbine usage is cut back to the optimum.

A third factor that tended to prevent efficiency improvements was the difficulty of designing boilers to operate at steam temperatures and pressures much higher than 1000°F and 3000 pounds per square inch. Many tons of alloy steel are required for fabricating the boiler tubes that contain the steam and outside of which the hot combustion gases flow, and the metallurgical limits on strength and corrosion resistance are being approached for alloys of reasonable cost. More highly alloyed boiler tubes would be required for higher-temperature steam and hence for higher-efficiency generation, but the cost of buying and fabricating them has so far outweighed the savings achievable through greater efficiency. Alternate technologies—in particular the gas turbine and steam turbine combined cycle discussed in the next chapter—are in process of overtaking the efficiency of the conventional steam plant, which may spur boiler designers to take the expensive step to more highly alloyed steels. But either or both ways new technology will push efficiency up beyond the plateau of the 1960's.

The technological basis for expecting the efficiency of electric power generation to improve in future decades is discussed in more detail in the next chapter. The results of that discussion have been anticipated here because of the key importance of improved efficiency to the future growth of electricity.

The substantial reductions in the price of electricity illustrated in Figure 11.4 were achieved in large part through economies of scale and new technology associated with the development of increasingly powerful boiler-turbine-generator combines. Individual power-plant capacity grew even more rapidly than electric power output, doubling every 7 or 8 years in recent decades, with the result that by 1970 the power plants with lowest unit cost had generating capacities in the neighborhood of 1000 megawatts (a megawatt is a million watts). Total United States generating capacity in 1970 was only 360,000 megawatts, so that just 360 of the most economic plants then available could have powered the country with electricity. In actuality, as shown in Table 11.1, the electric utility industry with 341,000 megawatts of the country's generating capacity operated approximately 2700 plants, most of them built long ago when the industry was much smaller.

An electric utility adding new generating capacity is reluctant to have the

TABLE 11.1 STRUCTURE AND GENERATING CAPACITY OF THE
UNITED STATES ELECTRIC UTILITY INDUSTRY (11.4)

Type of System	Number of Systems	Capacity in Thousands of Megawatts	Number of Plants
Generating systems			
Investor owned	~250	263	1923
Federal	2	39⎫	
Public nonfederal	~700	34⎬	~800
Cooperative	~65	5⎭	
	~1000	341	~2700
Nongenerating systems			
Investor owned	~150		
Federal	3		
Public nonfederal	~1360		
Cooperative	~890		
	~2400		

~ indicates approximate number.

new power plant represent more than at most 20 percent of its total capacity,
because unanticipated shutdown of a larger plant could put too severe a strain
on the rest of the system as it takes up the extra load. Hence, in order to take
full advantage of modern technology's potential for low unit cost, a utility
would have to have a system generating capacity of 5000 megawatts or more.
It is clear that the average electricity-generating utility with only about 340-
megawatt capacity is far too small to take advantage of the current economy of
scale. Only the 20 largest utilities exceed the 5000-megawatt system
generating capacity that would allow them to add a 1000-megawatt plant on
their own. Yet the situation is better than it might seem, for these large
utilities together account for half of the country's total generating capacity. In
addition, smaller utilities increasingly band together to share a modern, low-
unit-cost plant, with the additional mutual advantage of establishing system
interties. Nongenerating utilities, which tend to be the smallest utilities,
purchase power from their larger neighbors. These potentialities for sharing
will allow the electric utility industry to take advantage of further economies of
scale should optimum power plant size continue to increase.

In the early days of electricity when power plants were smaller, individual factories faced with a choice of purchasing electricity or generating their own often chose to generate their own. But as the size of the optimum power plant increased more rapidly than the power requirements of any single factory, industrial users found themselves in much the same situation as the small utilities, and they tended increasingly to purchase their electricity. As a result, the electric utility industry's share of the nation's generating capacity increased from 65 percent in 1920 to 95 percent in 1970 (11.2), and the generating capacity of the utility industry grew faster than that for the nation as a whole.

In analyzing the progress of electrification, three different energy quantities are of interest: (1) the energy input to electric power generation, (2) the electric energy output from electric power generation, and (3) the electric energy output from power generation by the electric utility industry. As efficiency of generation increases, electric energy output grows more rapidly than energy input to electric generation; and as the utility industry wins a larger share of total electric power generation, the utilities' electric power output grows more rapidly than the nation's total output. Recent growth rates for these three energy quantities in the United States are given in Table 11.2.

Overall energy input to electric generation grew at 6.1 percent annually during the period 1960–1968, compared with the 4.3 percent annual growth rate for total energy consumption. Over the same time span, the proportion of the nation's energy input turned into electricity increased from 20.8 percent to 25.0 percent, accounting for most of the differential growth of energy for electric power generation relative to total energy.

Growth rates for the various end uses of electricity during the period 1960 to 1968 are shown in Table 11.3, which classifies end uses by their degree of

TABLE 11.2 UNITED STATES ENERGY GROWTH RATES 1960–1968 (11.5)

Growth Entity	Annual Growth Rate	Explanation of Difference
Energy for electricity	6.1%	Improved efficiency: electric output increases per unit energy input
Electricity	6.7%	
Utility electricity	7.2%	Utilities pick up a larger share of total generation

TABLE 11.3 ELECTRIFICATION OF THE UNITED STATES 1960–1968
(11.6)

End Uses	1968 Percent Electric	1968 Percent of United States		Percent Annual Growth	
		Energy	Electricity	Total Energy	Energy for Electricity
Electrified					
Industrial drive	∼100	10.3	39.7	4.4	4.4
Refrigeration	∼100	3.0	11.6	5.6	6.0
Electrolytic processes	100	1.5	5.8	3.7	3.7
Air conditioning	96	1.5	5.6	11.3	11.1
Television	100	0.7	2.6	9.6	9.6
Lighting and other	∼100	5.0	19.1	9.4	9.4
	∼100	22.0	84.4	6.0	6.0
Partially electrified					
Clothes drying	70	0.4	1.0	10.2	9.9
Cooking	40	1.4	2.1	2.0	3.0
Water heating	38	4.2	6.2	4.1	3.4
Direct heat	6	11.0	2.6	2.8	5.8
Space heating	5	20.8	3.3	4.1	24.
	11	37.8	15.2	3.6	6.6
Unelectrified					
Transportation	—	25.6	0.4	4.1	—
Process steam	—	14.6	—	3.8	—
	—	40.2	0.4	4.0	—
All end uses	25	100	100	4.3	6.1

∼ indicates approximate percentage.

electrification. The first portion of the table is a repeat of Table 9.2 showing
the percentage consumptions of energy and electricity in 1968 for each end use.
(In going from Table 9.2 to Table 11.3, it was necessary to lump "lighting"
and "other" into a single category because data are not available for lighting
in 1960.) The last two columns show two annual growth rates for each end
use: one for total consumption of energy, and the other for consumption of the
energy that is transformed to electricity prior to use. These two growth rates
are essentially the same for the electrified end uses, as they ought to be. Slight

departures from equality for refrigeration and air conditioning reflect compensating trends wherein refrigeration moved a little closer to full electrification and air conditioning moved a little farther away from full electrification. Growth rates for the partially electrified end uses show increased electrification of cooking, direct heat, and space heating; and decreased electrification of clothes drying and water heating. Overall, largely because of increased electrification of space heating and direct heat, consumption of energy via electricity for the partially electrified end uses grew nearly twice as rapidly as did total consumption of energy for these uses, at 6.6 percent annually.

It is difficult to project future consumption of electricity without at the same time projecting (at least implicitly) the related technological improvements in efficiency of generation and use. Yet a partial perspective can be gained by projecting the growth of electric energy consumption by end use under different assumptions concerning the pace of technological progress, and lower and upper bounds on total future consumption of electricity can be estimated.

The electrified end uses are perhaps the easiest to project, because there is every likelihood that they will stay electrified. Insofar as technology improves the efficiencies of generation and use of electricity, the more surely will consumers be inclined to stick with electricity. Hence the consumption of energy via electricity will equal the total consumption of energy for these end uses, for which projections have already been made in Chapter 3 and are detailed in Appendix 3.

The unelectrified end uses also are easy to project, aside from the possibility that transportation may become electrified to a significant extent. However, for reasons discussed in previous chapters, I anticipate neither electrification of transportation nor electrification of process steam. Hence the unelectrified end uses are projected to stay unelectrified at least for the next few decades (although there remains an outside chance that a breakthrough in battery technology could result in the electrification of automobiles).

The partially electrified end uses are more difficult to project. The very fact that some consumers choose electricity while others choose direct combustion of fuel suggests that changes in relative convenience and cost to consumers that may result from advances in technology may significantly affect the course of electrification. Consider, for example, the electrification of space heating, which proceeded rapidly during the 1960's, albeit from a very small base and to only 5 percent penetration. Decisions on residential heating sometimes are made by builders, for whom the cost and convenience of construction and installation are of prime significance; they are sometimes made by owners, for whom in addition the operating costs, cleanliness, and convenience of thermostatic control are of significance; and they sometimes are affected by local availability or unavailability of natural gas for heating new residences. These various factors influenced decisions increasingly toward electric space heating during the 1960's. But to what extent is this trend and its future continuation or noncontinuation dependent on improvements in technology? Would a shortage of natural gas combined with convenience features of value to both builders and owners continue to increase the electrification of space heating in the absence of further improvement in electric-generation efficiency or heat-pump performance? Or would improvements in efficiency of generation and in performance of heat pumps be necessary if electrification of space heating is not to level off at a quarter or a third of new construction?

Perhaps the bleakest projection that could be made with respect to future electrification would be based on the assumptions of no technological improvement and no change in electricity's share of any partially electrified end use. With these assumptions, the growth in electric power consumption for each group of end uses would be just proportional to the growth in total energy consumption for that group of end uses. This calculation is carried out in Appendix 3. The resulting growth in energy input for electric generation amounts only to a factor of 2.24, as does the growth in electric energy output owing to the assumed lack of technological progress in improving conversion efficiency. This would represent a dramatic slowdown from the past history of a doubling of electric power output every decade. At the other extreme, perhaps the rosiest projection that could be made with respect to future electrification would assume that efficiency of generation and end use improved so dramatically that all end uses except process steam and transportation became fully electrified. If this were to happen by the year 2000, the calculation in Appendix 3 shows that energy input to the generation of electricity would increase by a factor of 4.6 between 1970 and 2000. The two extremes just considered are sketched in Figure 11.5. They constitute lower and upper bounds for the growth of energy input to electric generation over the next 30 years.

In order to achieve the full electrification envisioned for the upper bound in Figure 11.5, I estimate that an average generation efficiency of 50 percent would be required for plants using fossil fuels, a substantial improvement over the utilities' 32.5 percent of 1970 and indeed a substantial improvement over the 40 percent that represents 1970's best commercial practice. An average efficiency of 50 percent is technologically possible, although I will defer judgment on the time scale for achieving it. Were it to be achieved by 2000, as required for the 4.6-fold increase in energy input to electric generation, the consumption of electricity would increase about seven-fold, coming close to doubling every decade for three more decades. (The electric utilities could increase output eight-fold if they were to achieve 100 percent of electric power generation by taking over the residual of industrial generation.) The two extremes for electric power output are sketched in Figure 11.6. They constitute lower and upper bounds for the growth of electric power over the next 30 years. Also sketched in Figure 11.6 is an intermediate projection based on converting 45 percent of United States energy input to electricity by 2000 and on achieving an average fossil-fueled plant efficiency of 45 percent by the same year.

The bounds in Figures 11.5 and 11.6 are not rigid bounds, for a concerted mass movement toward conservation and against technology might pull actual performance below the lower projections, and my estimate of the future general level of energy consumption could be too low on any one of a number of counts so that actual performance could rise above the upper projections. The wide spread between lower and upper bounds shows the significance of

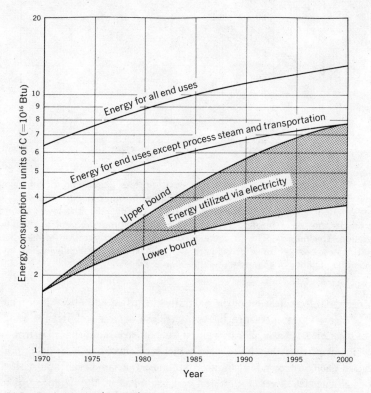

Figure 11.5. Projections of United States energy consumption, and of energy utilized via electricity (11.7). For energy utilized via electricity, the lower bound corresponds to no change in the degree of electrification of any end use. The upper bound corresponds to full electrification by 2000 of all end uses except process steam and transportation.

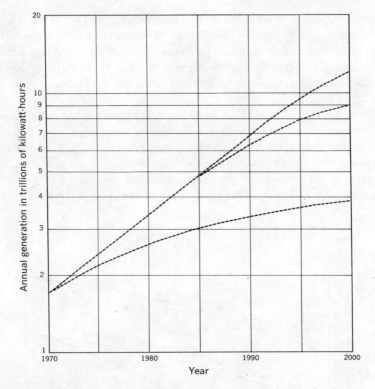

Figure 11.6.   Projections of United States electric power generation for the years 1970 to 2000. The upper bound corresponds to the upper bound of Figure 11.5 with average power generation efficiency advancing to 50 percent by 2000. The lower bound corresponds to the lower bound of Figure 11.5 with no improvement in efficiency of generation. The intermediate projection corresponds to conversion of 45 percent of United States energy input to electricity by 2000 and to achieving an average conversion efficiency of 45 percent by the same year.

the degree of electrification of the partially electrified end uses, all of which are uses of heat. Consumers are likely to choose between electricity and direct combustion on the basis of cost, at least for industrial and space heating, which tends to mean that the choice goes whichever way uses less fuel input. The marginal choice corresponds to approximately the same fuel input either way, and as efficiencies of generation and end use increase, the balance tips toward increased electrification.

The upper bounds can be achieved only through substantial improvements in generation efficiency, which means that electric energy output must increase faster than the input of energy to electric generation. This multiplying effect, where greater efficiency leads to more electric output per unit fuel input, on top of increased fuel shifted to electric generation, is reflected in the rapid growth of electric output projected for the upper bound in Figure 11.6. Yet the total consumption of fuel—direct combustion plus input to electric generation—is relatively unaffected.

CHAPTER TWELVE

ELECTRIC GENERATION

TECHNOLOGIES

Electricity can be generated through conversion of heat from fossil or nuclear fuel (as in a steam plant), of chemical energy (as in a fuel cell), or of solar energy (as in a hydroelectric plant or solar cell). Alternative technologies are available at various stages of development for each type of energy source, and it is useful to classify them by the degree of commercial acceptance they have so far achieved, as is done in Table 12.1. Major generation technologies are defined to be those that provide 1 percent or more of world electric generation capacity. Special situation technologies are commercially successful technologies that have carved out a small niche or market segment less than 1 percent of world generation and that seem unlikely to grow out of their niches. New technologies do not yet provide as much as 1 percent of world generation capacity, but are intended by their developers and promoters to do so. Laboratory demonstrations have been shown to produce electricity, through past or current research and development support, but do not yet have commercial backing. Theoretical possibilities, whether or not the subjects of research and development, have not yet reached the stage of successful demonstration in the laboratory.

Whatever the primary energy source, all major electric generation technologies are based on a turbine prime mover driving a rotating electric generator. (The small electric generators in automobiles and trucks actually have more aggregate generating capacity than the entire electric utility in-

TABLE 12.1   ELECTRIC GENERATION TECHNOLOGIES

Major generation technologies (prime mover and generator)
   Hydraulic turbine
   Fossil-fuel steam turbine
      Coal-fired boiler
      Natural-gas-fired boiler
      Oil-fired boiler
   Combustion gas turbine
   Nuclear steam turbine
      Boiling water reactor (BWR)
      Pressurized water reactor (PWR)

Special situation technologies
   Diesel engine and generator
   Geothermal steam turbine and generator
   Agricultural products steam turbine and generator
   Solar cells
   Batteries
   Fuel cells

New technologies (prime mover and generator)
   Nuclear steam turbine
      Advanced gas-cooled reactor (AGR)
      Heavy-water reactor
      High-temperature gas-cooled reactor (HTGR)
      Liquid-metal fast-breeder reactor (LMFBR)
   Combustion gas turbine and steam turbine combined cycle

Laboratory demonstrations
   Nuclear steam turbine and generator
      Molten-salt reactor
   Nuclear gas turbine and steam turbine combined cycle, & generator
   Thermionic generator
   Magnetohydrodynamic generator (MHD)
   Ocean water heat source and sink

Theoretical possibility
   Nuclear fusion heat source

---

dustry, although the amount of electricity they generate is relatively small because automotive generating capacity is idle a large fraction of the time and is seldom utilized to capacity when operating. However, aside from this parenthetical comment, I have neglected automotive electric power and generating capacity.) Turbine power can come from falling water, combustion gases, or steam generated from fossil or nuclear fuels. New technologies also are based on the turbine-generator combination. The only technologies listed in Table 12.1 not dependent upon prime movers driving generators are solar cells, batteries, and fuel cells among the special situations and thermionic and MHD generators among the laboratory demonstrations.

Prior to 1960 the generation of electricity was based largely on hydropower, fossil steam, and diesel power. Hydropower was used where available; fossil-fuel steam was the choice for large, efficient, low-unit-cost plants at other locations where demand was adequate to justify a large plant; and diesel power was utilized in more remote locations with smaller demand. Then during the 1960's, additional technologies began making inroads into the commercial generation of electricity in competition with the traditional fossil-fuel steam plant. Nuclear technologies, which generate heat through nuclear fission rather than through fuel combustion, and gas turbine technology, which eliminates boilers, heat transfer surfaces, and steam by passing the products of combustion directly through the turbine, became well established. And there is an ever-growing list of additional candidates for commercial electric generation.

The technologies I judge to have current commercial significance, together with selected other technologies of general interest, are listed in Table 12.1. Because the pace and extent of electrification depend so strongly on the efficiency of generation and on other technological advances that may lead to future economies of scale, it is important to review these technologies and their probable future development. Most of the remainder of this chapter is devoted to a brief preliminary review of this type, one paragraph to each technology, including a preliminary estimate of its probable future commercial significance. For the most promising technologies, these estimates will be substantially refined in the next two chapters. Key factors for assessing commercial significance include capital cost of generating capacity and efficiency of generation. High capital cost is an unfavorable factor, which can eliminate a technology even if its energy source is free; and low efficiency is an unfavorable factor, which can eliminate a technology even if its plant cost is negligible. Potential commercial winners must combine low capital cost with high efficiency.

Hydroelectric power technology is old and unchanging. Flowing water is dammed and diverted through water-powered turbines of high efficiency and low speed that are connected to low-speed generators. The technique has small potentiality, because of the limited availability of falling water. Already perhaps half of the country's potential has been tapped. Pumped-storage hydroelectric systems may play a small role in peaking power, but conservationists are increasingly opposed to new dams and inundations.

Fossil-fuel steam power technology is well developed. Fossil fuel is burned at near-atmospheric pressure, and a portion of the heat is transferred in a boiler to steam at high temperature and pressure. The steam is passed through a steam turbine, condensed, and recycled. The turbine turns a direct-connected generator. After rising to 1050°F in the mid-1960's, steam temperature has stabilized at about 1000°F. Modern steam plants with 1000°F steam can achieve an overall thermal efficiency close to 40 percent compared to a theoretical limit of about 60 percent for a perfect heat engine working between the same temperature limits. Further increases in efficiency will not be easy, because alloys of substantially greater cost are necessary if higher steam temperatures are to be employed. Boilers designed to handle coal are more expensive than those designed solely for gas or oil, owing to the corrosive and erosive properties of hot coal ash and other combustion products.

Combustion gas turbine technology shows great potential for further development. The basic gas turbine consists of a compressor, which draws in air and raises its pressure, a combustion chamber where fuel is added to the compressed air and burned, and a turbine where the hot combustion gases give up some of their energy as work. The turbine drives the compressor, and the excess work is available for driving a generator. Until rotary gas compressors were developed in the 1940's for aircraft gas turbine applications, gas turbines were not available at all. Then as the axial-flow compressor was introduced, as compressor design and turbine design improved, and as better materials were developed to withstand high temperatures, efficiency improved. Future improvements in materials, turbine bucket cooling, and high-performance compressors can increase thermal efficiency of the simple gas turbine substantially beyond the present 30 percent. The cost penalty of exotic materials is not so severe for gas turbines because of the relatively small amount of material required, there being no boiler tubes, no valves, and a thinner turbine casing because of the lower working pressure.

Nuclear steam technologies are the most complex of the major technologies. Light-water reactors, which account for practically all plants built or under construction in the United States, utilize the uranium isotope U235 as fuel. The U235, enriched to a few percent concentration in admixture with the more common U238 isotope, is packaged in fabricated zirconium-encased fuel elements, which in turn are supported in a three-dimensional array through which primary water can flow and be heated. When the geometrical configuration is correct, controlled fission occurs in the U235 atoms in the array of fuel elements. The energy released in fission heats the fuel elements which in turn heat the primary water. In the boiling water reactor (BWR), the primary water boils and steam is produced. The steam then passes to a steam turbine, is condensed, and is recycled, as in a conventional steam plant. In the pressurized water reactor (PWR), the primary water does not boil, but is maintained in the liquid state under pressure. The pressurized primary water flows through a heat exchanger where it heats and boils secondary water (primary and secondary water do not mix because they are separated by a metal barrier in the heat exchanger). The primary water returns to be reheated in the reactor, and the steam from the boiling secondary water passes through a steam turbine and is recycled, as in a conventional steam plant. The promise of nuclear steam generation is the availability and low cost of fuel per Btu of energy. The problem of containment of large volumes of high-pressure steam or pressurized water tends to limit the steam temperature to about 550°F, which reduces system efficiency and adds to the size and cost of turbine and heat-recovery equipment per kilowatt of electrical output. All nuclear technologies must overcome problems associated with nuclear radiation, which damages structural materials and fuel elements, makes necessary special shielding to protect living organisms, and adds to the cost of fuel reprocessing where fissionable elements are reclaimed from partially spent fuel.

Diesel engine technology is highly developed and is utilized primarily for small installations or for peaking power where low capital cost and ability for rapid start-up are of particular value.

Geothermal steam is available in limited amounts at a few places where natural steam flows from wells. It has not proven feasible to utilize geothermal heat other than by the discovery and exploitation of clean steam wells. For the future, geothermal energy is likely to be limited to a few special situations of overall minor significance to electric generation.

Combustion of agricultural products for steam generation in the United States is limited largely to scrap wood and sawdust. The technology is identical with fossil-fuel steam, and is an alternative for fossil-fuel steam where agricultural products are cheap.

Solar cell technology is based on solid-state photocells, of the type used in exposure meters and space vehicles, which convert about 10 percent of the energy of sunlight that falls on them directly to electric energy. Although solar cells are firmly entrenched in several special applications, their cost is much too high for major commercial electric generation.

Batteries convert chemical energy directly to electric energy without the intervention of a heat engine. Hence battery efficiency is not limited to the theoretical heat engine efficiency, and it can approach 80 to 90 percent. There are two general classes of batteries: primary batteries, such as the ordinary flashlight dry cells in which the chemicals react once to produce electricity, after which the battery is discarded; and storage batteries, such as the automobile lead-acid battery in which the chemical reactants can be regenerated by a reverse flow of electricity from the outside. Batteries have many applications, but their cost is too high for them to become a major factor in electric power generation. Much battery research is directed toward discovering a battery suitable for automobile propulsion in competition with the internal combustion engine. If successful, this development would not constitute a new form of electric power generation, but rather a new form of electric energy storage.

Fuel cells transform chemical energy directly to electric energy in a manner similar to that in batteries, except that the chemicals that produce the electricity—fuel and air or oxygen—are continuously fed to the fuel cell, and the waste products of the electricity-producing reaction are continuously removed. Direct-current electricity of about 1 volt is produced at high efficiency, in the neighborhood of 70 percent, with a theoretical limit approaching 100 percent. Early fuel cells designed for spacecraft required hydrogen and oxygen as reactants and platinum as a catalyst, and as a result were much too expensive for electric utility power generation. Current development effort is focused on high-temperature fuel cells utilizing air and gasified coal or air and natural gas as reactants.

The United Kingdom's advanced gas-cooled reactor (AGR) utilizes uranium fuel and substitutes carbon dioxide gas for water as the primary fluid that circulates through the reactor and is heated by the fuel elements. The fuel elements are encased in stainless steel instead of zirconium. The primary fluid delivers heat to a secondary loop of water which is converted to high-temperature steam at about 1000°F, so that greater overall plant efficiency can be achieved. AGR plants are just coming into commercial operation.

Canada's heavy-water reactors utilize a special isotopic form of water, deuterium (heavy hydrogen) oxide D_2O instead of H_2O, as neutron moderator and as primary pressurized-water coolant. The nuclear properties of D_2O make possible the use of the U235 contained in natural unenriched uranium as fuel.

A high-temperature gas-cooled reactor (HTGR) is similar to an AGR, except that helium is used instead of carbon dioxide as the primary fluid that picks up heat from graphite-coated fuel elements. By utilizing inert helium gas as working fluid, problems of corrosion, radiation damage, and radioactivity can be minimized, and a higher working temperature and greater efficiency can be achieved. HTGR plants are coming into commercial operation and show high promise of becoming a major electric generation technology.

Breeder reactor technology is able to create (breed) more fissionable fuel than is consumed in the power reactor. In a liquid-metal-cooled fast-breeder reactor (LMFBR), the common uranium isotope U238 is placed in and around the reactor where it can absorb some of the excess fast neutrons generated by fission of the fuel. Under the influence of fast neutrons, U238 passes through a series of nuclear reactions and is transmuted to plutonium 239, which is a fissionable nuclear fuel. In a well-designed breeder, more plutonium is created than fissionable fuel is consumed, with the result that most natural uranium can ultimately be transmuted to plutonium and utilized as fuel. Liquid sodium metal circulates as a primary coolant through the reactor and gives up its heat to a secondary flow of water which is converted thereby to steam for powering a steam turbine. In addition to the technological problems associated with light-water slow-neutron reactors, the LMFBR must contend with severe metallurgical problems associated with molten sodium and with the structural materials damage caused by fast neutrons and larger percentage fuel burnup. Pilot plant LMFBR's are in operation around the world, and substantial worldwide effort is being devoted to making them commercial. Molten-salt reactors, described in the paragraph after next, represent another branch of breeder technology.

Gas turbine and steam turbine combined-cycle technology has promise for high efficiency with low capital cost. The gas turbine makes possible a very high upper working temperature T_1 compared with the temperature of the steam in a conventional fossil-fuel boiler. By passing the hot turbine exhaust through a heat recovery boiler, steam is generated for operating a steam turbine, which makes possible a lower working temperature T_2 compared with the exhaust temperature of a simple gas turbine. Hence in combination the gas turbine and heat recovery steam turbine can achieve an overall higher efficiency than either alone. Commercial combined-cycle plants are now available with efficiencies as high as conventional fossil-fuel steam plants, and combined cycle technology is progressing rapidly.

Molten-salt reactors have been constructed with thorium instead of U238 as the fertile material. Fuel salts, fertile-thorium salts, and carrier salts are melted together to form a hot fluid that circulates through a graphite reactor core. Nuclear reaction heats the fluid and transforms some of the thorium to U233, a fissionable isotope of uranium not found in nature which can serve as fuel. The hot salt is utilized to heat a working fluid such as steam, and portions of the salt are cycled through continuous chemical-recovery loops to remove excess U233 fuel that has been bred and to remove undesirable fission products that otherwise would poison the reactor by absorbing and wasting neutrons. The molten-salt reactor is a chemical engineering approach to breeding; whereas the sodium-cooled fast breeder is a mechanical engineering approach. Molten-salt technology has not progressed beyond the construction and operation of two small experimental reactors.

Nuclear gas turbine and steam turbine combined-cycle technology offers a method of improving on the simple HTGR. The inert helium gas in the primary circuit is compressed to high pressure, then circulated through the nuclear reactor where it is heated to high temperature, then allowed to expand through a gas turbine where it delivers work for gas compression and electric generation, then passed through a heat recovery boiler where residual heat is removed and steam generated, then is compressed again to high pressure and recirculated. The helium gas turbine and intermediate-temperature heat recovery boiler in combination are expected to be more efficient and less costly than the high-temperature heat exchanger in the HTGR.

Thermionic generation is based on the fact that a high-temperature metal surface emits electrons in very much the same way as a high-temperature water surface emits steam. The electrons boiled off a hot metal surface can be collected on a nearby cooler metal surface at higher voltage, from which they are not so strongly emitted because of the lower temperature, and can be caused to flow through an external circuit in returning to the original emitter. Unfortunately, space-charge effects require the two surfaces to be placed very close together (less than a thousandth of an inch), and it has not been possible to develop thermionic generators beyond the laboratory demonstration stage.

MHD is an acronym for magnetohydrodynamics. MHD technology seeks to utilize the hot high-velocity gas stream from a stationary rocket engine as the moving conductor in a direct-current generator. In a conventional generator, a metal conductor is forced through a magnetic field, and a voltage is induced in the conductor. When the conductor is connected to an external load, current flows and electrical power is generated. In an MHD generator, an electrically conducting rocket-exhaust gas stream is forced (by its own inertia) through a magnetic field, and a voltage is induced across the gas. Electrodes on either side of the gas stream pick up the voltage and connect to an external load through which current then flows. The major technological problems of MHD are the electrical conductivity of the hot gas, which is too low, and the rapid erosion of the electrodes in the hot gas stream.

In many tropical oceans, the surface water is substantially warmer than the deep water underneath. It is possible to boil a liquid such as ammonia using the warm water as heat source, to run the vapor through a vapor turbine that drives a generator, and to condense the vapor back to liquid using the cold water as heat sink. High capital cost and vulnerability to barnacles and storm damage may be the most important reasons for lack of interest in this technology.

Nuclear fusion technology is not yet developed. The idea is to fuse tritium or deuterium atoms (heavy isotopes of hydrogen) into helium atoms at a controlled rate, thereby liberating the vast quantities of energy that are liberated more rapidly and (for most purposes) less usefully in large thermonuclear explosions. The fusion process requires very high temperatures, in the neighborhood of 100 million degrees. Such temperatures can exist where the hot material is confined by nonmaterial walls such as provided by a magnetic field, and they can exist transiently while the hot material is exploding. Experiments with magnetic confinement have produced fusion, but less energy than was consumed. Experiments with tiny thermonuclear explosions set off by pulses of laser radiation have produced fusion, but less energy than was consumed. I expect that with further development, both methods will generate more energy than was consumed, and these processes will surely have applications. However, there is no chance at all that they can be utilized in this century for the generation of electricity at a cost competitive with other technologies.

Each of the technologies briefly reviewed in this chapter has advocates and champions, whose efforts are directed toward promoting the status of a theoretical possibility to a laboratory demonstration, or a laboratory demonstration to a special situation or new technology, or a special situation or new technology to a major technology. They cannot all succeed. Overall, the technologies most likely to succeed at future major electric power generation are descendants of today's fossil-fuel steam, nuclear steam, and gas turbine technologies, all of which are capable of further development enabling further reduction in the cost of electricity. In summary, the most promising technologies for future electric power generation are

1. Fossil-fuel steam with higher efficiency through higher upper working temperature T_1
   a. via higher-temperature boiler
   b. via gas turbine combined cycle
2. Nuclear steam with lower capital cost per unit of electrical output
   a. via standardization and faster licensing of BWR and PWR reactors
   b. via high-temperature, high-efficiency reactors such as HTGR, LMFBR, and nuclear gas turbine combined cycle
3. Gas turbine with lower cost and higher efficiency
   a. for intermittent peaking applications
   b. for use in combined cycles.

In addition to prime mover technologies summarized above, other technologies are of interest and importance to electrification. They include electric generators, transformers, transmission lines, cables, energy storage for peaking generation, heavy-duty fuel cells, and heat pumps. Potential progress in these technologies is discussed at appropriate places elsewhere in the book.

CHAPTER THIRTEEN

FACTORS AFFECTING

THE COST OF ELECTRICITY

The future course of evolution of the electric utility industry is strongly dependent on the costs of fuel, capital equipment, and operations. The fuels and conversion systems that will win out in the end must bring electricity to the consumer at the lowest price, while satisfying socially and politically motivated regulatory constraints. Advances in technology have played key roles in the past, and are certain to do so in the future. Yet the various areas of technology will not prove to be of equal value. As in the past, only a few will meet the rigorous test of competition.

At the present time, nuclear fuels are candidates for replacing fossil fuels, and the gas turbine is a candidate for replacing high-temperature boilers. Many other technologies are being advocated as possible candidates for the more distant future. These new technologies are all capable in principle of producing electricity, but many will never have the capability of doing so competitively. Significant effort, if devoted to them, would be wasted. It is important to forecast the future of technology in relationship to the basic social, political, and economic factors of the electric utility industry, if the key areas of technology—those whose vigorous pursuit will bring into being the electric energy conversion systems of the future—are to be identified and supported today.

The major elements of cost in the electric utility industry are:

Cost of fuel
Cost of capital
   for generation equipment
   for transmission equipment
   for distribution equipment
Cost of operations and maintenance
Cost of transmission losses

Together they determine the cost of electricity. Economic comparison of alternate techniques for supplying electric energy requires that the cost of electricity be determined for whatever numerical values may be assigned to these cost elements. The numerical values depend on many factors, for example, the type of power plant under consideration, and they may change with time under the influence of future technology, economies of scale, fuel availability, and altered regulations.

In comparing the costs of electricity for alternate generation and transmission technologies, a common measuring point must be established. When the alternatives to be compared share the same transmission and distribution facilities, the power plant is a convenient measuring point, and a comparison can be made in terms of the cost of electricity at the power plant. When the alternatives involve different transmission facilities, then the substation, where high voltage transmission ends and intermediate voltage distribution begins, provides a convenient measuring point, and comparison can be made in terms of the cost of electricity at the substation. The additional cost of distribution is common to all examples under consideration; hence it cannot affect the comparisons.

The cost of fuel is the simplest cost factor to analyze. Fuel cost for electric utility generation is usually measured in cents per million Btu, and representative costs for fossil fuels in the United States during the 1960's (measured in 1970 dollars) were:

Fuel	Cost (delivered)
Coal	30¢/million Btu
Natural gas	30¢/million Btu
Oil	40¢/million Btu

Coal was consumed primarily near mines and population centers in the Midwest and East, natural gas near wells in the Southwest, and oil near ports in the Northeast. Most of the coal and oil contained a few percentage points of sulfur, as environmental sulfur restrictions had not yet been enacted. The natural gas cost was low because of FPC price regulation.

If the energy of fuel combustion were completely converted to electric energy with 100 percent efficiency, energy in the amount of 3412 Btu would be required to generate each kilowatt-hour (kWh) of electric energy. But because generation efficiency is less than 100 percent, the number of Btu required to generate a kilowatt-hour is larger than 3412. A steam plant operating at 40 percent efficiency would require 8530 Btu of heat input per kilowatt-hour of electric output. This ratio of Btu per kilowatt-hour is called the "heat rate," and one of the goals of power plant engineers is to progressively reduce the heat rate toward the theoretical minimum of 3412 Btu per kilowatt-hour every time they design a new power plant.

The cost of fuel in cents per million Btu, combined with the heat rate in Btu per killowatt-hour, together determine the cost of fuel per kilowatt-hour. As an example: coal at 30¢ per million Btu in a steam plant with a heat rate of 10,000 Btu per kilowatt-hour gives

$$\left(\frac{30¢}{1,000,000 \text{ Btu}}\right) \times \left(\frac{10,000 \text{ Btu}}{1 \text{ kWh}}\right) = 0.3¢/\text{kWh}.$$

It is the custom in the utility industry to express costs per kilowatt-hour in terms of mills, 1 mill being a tenth of a cent,

$$0.3¢/\text{kWh} = 3 \text{ mills/kWh}.$$

The units of measurement used here are everyday aids to thought and action for electric utility engineers and economists, although they may be less familiar to others. They will be used in all examples and comparisons, and readers are encouraged to adjust to them:

Cost element	Conventional unit
Cost of fuel	¢/million Btu
Heat rate	Btu/kWh
Cost per unit electric output	mills/kWh

The approach in this chapter is to express major factors that contribute to the cost of electricity in terms of mills per kilowatt-hour. The total cost of electricity is the sum of the separate contributions.

The contribution of fuel cost to electricity cost is directly proportional to the product of (1) fuel cost per unit heating value and (2) heat rate. If the cost of fuel is doubled the contribution of fuel to the cost of electricity is doubled. If the heat rate is halved through dramatic improvement in power plant effi-

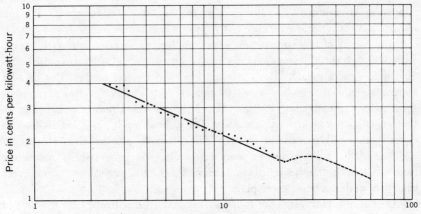

Cumulative production in units of 10^{12} kwh

Figure 13.1. Experience curve for the production of electric power by the United States electric utility industry 1943 to 1972 (13.1). Data previously shown in Figure 11.4 are extended to include the years 1971 and 1972 in which the price of electricity began to rise as a result of the shift from high-sulfur fuel to more costly low-sulfur fuel. If the cost of fossil fuel should double by the time the switch is complete, and if alternatives such as stack gas cleaning should prove to be as expensive as switching fuels, the price of electricity will tend to increase about 20 percent before taking up a new downward trajectory parallel to the old one.

ciency, the contribution of fuel to the cost of electricity is halved. As an example, compare the contribution of fuel to the cost of electricity from a Midwest coal-burning steam plant before and after sulfur emission restrictions were imposed. Sulfur emissions due to combustion of high-sulfur coal can be controlled by installing equipment to remove sulfur oxides from the stack gases after combustion or by switching to low-sulfur coal. Some utilities have gone one way and some the other, and in this example we consider the effects of a switch to low-sulfur coal from the Western states. This coal is cheaper at the mine but more expensive to transport because it must travel so far. The following numerical values are illustrative only, and do not necessarily correspond to any actual mines, railroads, or power plants.

Cost Element	High-Sulfur Coal	Low-Sulfur Coal
Cost at mine	\$3.60/ton	\$1.20/ton
Cost of transportation	\$3.60/ton	\$12.00/ton
Cost at power plant	\$7.20/ton	\$13.20/ton
Heating value	12,000 Btu/lb	11,000 Btu/lb
Cost per million Btu	30¢/million Btu	60¢/million Btu

Overall the cost of fuel at the power plant is doubled, and the fuel contribution to the cost of electricity also is doubled. If the plant has a heat rate of 10,000 Btu per killowatt-hour, typical of existing fossil-fuel steam plants, the fuel cost rises from 3 mills per kilowatt-hour to 6 mills per kilowatt-hour, an increase of 3 mills per kilowatt-hour. This increase amounts to something like a 20 percent increase in the cost of electricity to the consumer. If we look back at Figure 6.4 we note there, consistent with this illustrative example, a projected doubling in the cost of utility coal followed by a steady decline as experience accumulates with respect to the new coal and the new transportation network. Had Figure 6.4 addressed utility oil instead of utility coal, the picture would have been much the same. When fuel cost rises the cost of electricity rises, and the increased cost tends shortly to be reflected in increased price. The influence of rising fossil fuel cost on the price of electric power is already apparent, as shown in Figure 13.1, which reviews the historic record through 1972 and projects the price of electricity ahead to a future in which utility fossil fuel cost may be doubled (and the price of electricity may be raised 20 percent) by the shift to clean low-sulfur fuel.

The net effects of switching to low-sulfur fuel are to reduce the emission of sulfur into the air (the desired environmental effect) and to increase the cost of electricity (the associated economic effect). Although the cost of electricity is increased, the social cost of atmospheric pollution is reduced. If the antipollution regulations have been properly conceived, the reduction in social cost will exceed the increase in electric power cost, and society will overall be better off. This is an example of "internalizing" social costs, where widespread social costs of an activity are transferred in altered form to those who benefit from the activity (in this case users of electricity) to achieve a fairer apportionment of costs and an overall reduction in the total cost to society.

If all utilities burning high-sulfur fuel were to behave as in the example just reviewed, the overall cost of electricity in the United States might increase by about $2 billion annually, an amount judged by some to be less than the annual social cost of sulfur emission in the 1960's. But the true cost increase is likely to be less than $2 billion annually, because many utilities will find low-sulfur fuel closer to home or will install less expensive equipment for sulfur removal from stack gases. Whatever the initial cost may be, it is likely to be whittled away over the years as electric power engineers, caught unprepared by the environmental movement, improve on their initial makeshift solutions.

The influence of heat rate on the cost of electricity can be illustrated by comparing two power plants of slightly different efficiency. Consider a modern power plant with 40 percent efficiency in converting the energy of fossil fuel to electric energy, and an improved power plant with 41 percent efficiency, as they might compare in the late 1970's when the cost of fossil fuel is likely to be

50¢ per million Btu or more. (Sulfur emission restrictions affect utility oil as well as utility coal, necessitating the use of more-expensive, low-sulfur oil; and FPC relaxation of natural gas wellhead price limitations is likely to allow the price of natural gas to rise.)

Cost Element	40% Efficient Plant	41% Efficient Plant
Heat rate	8530 Btu/kWh	8322 Btu/kWh
Fuel cost	50¢/million Btu	50¢/million Btu
Fuel cost/kWh	4.26 mills/kWh	4.16 mills/kWh

The calculation shows that an improvement in efficiency of 1 percentage point from 40 percent to 41 percent cuts the cost of electricity by 0.10 mill per kilowatt-hour, or by about one-half of 1 percent of total electricity cost. In terms of the nation's electric bill, a saving of 0.1 mill per kilowatt-hour would amount to about $150 million annually. This is a substantial motivation for increasing plant efficiency, but as we see when capital costs are considered, the saving will be lost if the more efficient plant is too costly. Still, there is every reason to believe that the optimally designed plant will be more efficient, for today's plants were designed when fuel was available at half the price and the potential saving was only half as great.

Now we move to consideration of the capital costs associated with electric power generation. When a utility decides to build a new nuclear power plant at a cost of several hundred million dollars, it cannot reach into its pocket and pull out the cash, for there isn't any there. It borrows most of the money from banks and other lenders, and gets the rest from investors by selling stock or more commonly by holding back some of the investors' earnings instead of paying all the earnings out as dividends. The lenders will not lend unless the utility pays interest on the loan and gradually repays the principal. The investors will not invest unless they too see that they will be repaid with increased future dividends to compensate for current dividends lost. Hence the act of assembling the money for a new plant carries with it the inevitable consequential duties of paying interest and repaying principal. And commercial operation of the new plant brings the inevitable consequence of annual property taxes. These inevitable annual costs associated with capital investment are called fixed charges. They vary from region to region and from time to time depending on tax rates and interest rates, but usually fall in the range of 12 to 17 percent of invested capital. A fixed charge rate of 15 percent will serve for illustrative purposes.

A modern 1000-megawatt nuclear plant costs about $280 million (measured in 1970 dollars), and fixed charges at 15 percent amount to about $42 million annually. On a per kilowatt of capacity basis, the capital cost is $280 per kilowatt, and the annual fixed charges are $42 per kilowatt. Since there are 8760 hours in a year, the maximum possible electrical output from a kilowatt of capacity would be 8760 kilowatt-hours per year. There is every motivation to operate an expensive plant all the time to keep the revenue from sales coming in, but necessary maintenance combined with other operating factors generally limits operation to something closer to 7000 hours per year, which amounts to 80 percent availability. Annual fixed charges of $42 per kilowatt spread over 7000 hours of generation works out to 6 mills per kilowatt-hour.

Reliability and availability of large power plants are factors with significant influence on the cost of electricity. Suppose that improved technology were to shorten the time required for routine maintenance and to diminish the frequency and duration of unplanned shutdowns so that availability increased from 7000 hours to 7880 hours annually (80 percent availability increased to 90 percent). Then the same fixed charges of $42 per kilowatt-hour would be spread over 7880 hours, leading to fixed charges of 5.33 mills per kilowatt-hour, a reduction of 0.67 mills per kilowatt-hour. This is a powerful motivation for power plant engineers to increase plant availability.

It takes a long time to build a nuclear power plant, and the money that is needed for construction is not needed all at once. It is borrowed a little at a time, some as long as 7 years prior to commercial operation, but the average dollar is borrowed about 2 years prior to commercial operation. Two years' interest on $240 million borrowed for plant construction amounts to about $40 million. The utility ordinarily borrows the additional $40 million to cover these interest payments and adds the additional borrowed money to the cost of the plant, increasing capitalization to $280 million. This "interest during construction" is a significant portion of the capital cost of nuclear power plant construction.

Licensing delays can add substantially to power plant cost, for interest must still be paid while a completed plant is waiting for an operating license. A delay of 1 year for a completed $280 million plant adds about $22 million more of interest during construction to its capitalization, making it a $302 million plant. The additional $22 per kilowatt capitalization adds about 0.3 mill per kilowatt-hour to the cost of electricity. On average, I estimate that environmentally motivated licensing delays have added about a third of a mill per kilowatt-hour to the cost of nuclear electricity, and that construction delays from other causes have added another third of a mill. These are very rough estimates, as it is not easy to isolate and quantify the various causes of delay in bringing nuclear plants to commercial operation, but they show the significance of delay.

The capital cost of transmission is of interest for two entirely different reasons. First, because of the possibility that environmental pressures may require more high-voltage transmission to go underground instead of overhead, and second, because some generation technologies such as gas turbines and fuel cells can be sited at substations where they do not require transmission. Transmission capital costs have increased gradually with respect to generation capital costs over the past several decades, although with wide fluctuations. At present they amount on average to about half of generation capital costs. Fixed charges on transmission capital need not be the same as for generation capital, but I continue to assume 15 percent for illustrative purposes. Consider an example where a newly planned plant is required to shift to underground transmission from planned overhead transmission. Overhead transmission is the least expensive way of transmitting electric power because the air serves as natural electric insulation and because air convection carries away the heat from transmission losses. Overhead wires are bare metal, but when it becomes necessary to put the wires underground, they must be carefully wrapped with protective insulation and armored against external damage. They cannot carry as much power without overheating because earth does not carry heat away as effectively as air convection. For these reasons the insulated and armored cable that is buried in the ground may cost 10 times as much as the overhead wire and supporting towers that it replaces. Here is the illustrative comparison:

Cost Factor	Overhead	Underground
Capital cost of transmission, remote nuclear plant site to city substation	$200/kW	$2000/kW
Cost per unit of electric energy delivered, based on 100 percent availability	2.28 mills/kWh	22.8 mills/kWh

The cost of shifting to underground transmission would be so high, more than doubling the total cost of electricity, that it would surely lead to relocation of the plant to a site much nearer the consumers.

The demand for electricity is not uniform through the day, or from day to day. Figure 13.2 shows schematically how demand can rise and fall on a summer day in an Eastern city. The area labeled "base load" represents that portion of demand that requires steady generation through the 24-hour period. The area labeled "intermediate load" represents the smaller portion of demand that requires generation for about half of each day. The area labeled "peak" represents the very small portion of demand that requires generation for only a few hours each day. The electric utility serving the city must have generating capacity available to meet the peak load demand as well as the base and intermediate load demands. Although fuel is consumed only when generating equipment is in operation, fixed charges continue on regardless of the number of hours per day the equipment operates. When fuel cost and capital cost are considered jointly and added together for various types of generating equipment, to determine the type that would minimize overall costs, different answers are obtained for base load generation and peak load generation.

Consider three generating plants typical of those on line in the early 1970's with the following illustrative costs:

Type of Plant	Capital Cost	Heat Rate	Fuel Cost
Nuclear steam	$280/kW	10,000 Btu/kWh	15¢/million Btu
Fossil-fuel steam	$210/kW	9,000 Btu/kWh	40¢/million Btu
Gas turbine	$120/kW	13,000 Btu/kWh	70¢/million Btu

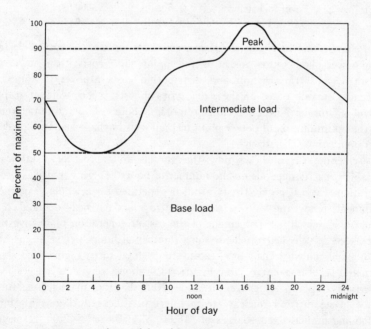

Figure 13.2. Variation in demand for electricity on a summer day in an eastern city (schematic). Base load demand requires steady generation throughout the 24-hour period. Intermediate load demand requires generation about half of each day. Peak demand requires generation for only a few hours each day.

For a full year's operation, the fixed charges for the above plants (at 15 percent) are

Type of Plant	Annual Fixed Charges per Kilowatt of Capacity
Nuclear steam	$42.00
Fossil-fuel steam	$31.50
Gas turbine	$18.00

Fuel costs per hour of operation or per thousand hours of operation are

Type of Plant	Operating Fuel Cost per Kilowatt of Capacity	
	per hr	per 1000 hr
Nuclear steam	1.5 mills	$1.50
Fossil-fuel steam	3.6 mills	$3.60
Gas turbine	9.1 mills	$9.10

Fixed charges and fuel costs together account for most of the cost of electricity at the power plant. Other operating and maintenance costs at the power plant are small, and transmission and distribution costs are yet to come. The "screening curve," based on these two types of costs, is a convenient graphical method of considering the full range of possible hours of operation per year, so that the optimum mix of power plant technologies for base load and peak load generation can be visualized.

A screening curve shows how the annual dollar cost of a kilowatt of available capacity depends on the number of hours per year that it is used for generation. Even if no electricity at all is generated for a whole year, the idle equipment, ready and waiting to respond to a sudden peak demand, has an annual cost, which is a percentage of the cost of generation plant investment. When electricity is generated for some number of hours per year, the cost of fuel must be added to the cost of capital to get total costs. Figure 13.3 shows a schematic screening curve, and Figure 13.4 shows screening curves for the three types of plants under consideration. Note that the curves for gas turbine, fossil-fuel steam, and nuclear steam intercept the vertical axis at the corresponding annual fixed charges of $18.00, $31.50, and $42.00, respectively. At zero annual hours of generation that is what they would cost. Note that at 1000 hours of operation the total cost is increased by $9.10 for gas turbine, $4.50 for fossil steam, and $1.50 for nuclear steam, as required for 1000 hours of operation, and that every additional 1000 hours of operation adds equal fuel cost increments. The figure shows that the gas turbine is least expensive for meeting peak load demands that aggregate less than about 2500 hours per year, that the nuclear steam plant is least expensive for base load demands that aggregate more than about 5000 hours per year, and that the fossil-fuel steam plant is least expensive for intermediate load demands in the range 2500–5000 hours per year. The lowest overall cost of electricity would be achieved by a utility whose generating system had some of each type of power plant: nuclear steam for base load, gas turbine for peak load, and fossil-fuel steam for intermediate load.

This general picture motivated the purchase of nuclear capacity and legitimized the purchase of gas turbines. Both types of plants were able to reduce the overall cost of electricity. Although the comparison in Figure 13.3 is typical of some utility systems in the early 1970's, it is not typical of all. It may not be typical of any in the 1980's and beyond as the costs of different generation technologies and fuels shift relative to each other. This question will be addressed in the following chapter.

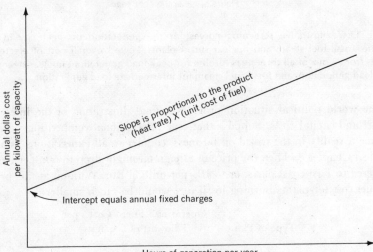

Figure 13.3.   Screening curve (schematic). A screening curve displays the annual dollar cost of a kilowatt of available capacity, as it depends on the number of hours per year that it is used for generation. Even if no electricity at all is generated for a whole year, the idle equipment, ready and waiting to respond to a sudden peak demand, has an annual cost equal to the fixed charges on the capital investiment. When electricity is generated for some number of hours per year, the cost of fuel must be added to the cost of capital to get total costs.

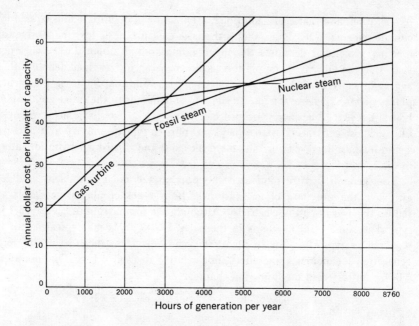

Figure 13.4. Illustrative screening curves for the generation of electricity in gas turbine, fossil-fuel steam and nuclear steam plants. Lowest overall cost of electricity results from a mix of all three; gas turbine for peak load generation, nuclear steam for base load generation, and fossil-fuel steam for intermediate load generation.

The world political situation provides a final illustration of the interplay of fuel and capital costs. Suppose that true international competition were to become a reality in the world oil business, counter to all expectations as analyzed in Chapter 8. Then the price of foreign oil might drop to 60¢ per barrel delivered to East coast ports, or to 10¢ per million Btu. With oil at this price, the fuel cost for plants burning fossil fuel would be much smaller:

Type of Plant	Operating Fuel Cost per Kilowatt of Capacity	
	per hr	per 1000 hr
Fossil-fuel steam	0.9 mills	$0.90
Gas turbine	1.3 mills	$1.30

The screening curves would shift as shown in Figure 13.5.

With oil so cheap, efficiency of generation would be less important, and the gas turbine would provide cheapest electricity for every type of load. From this

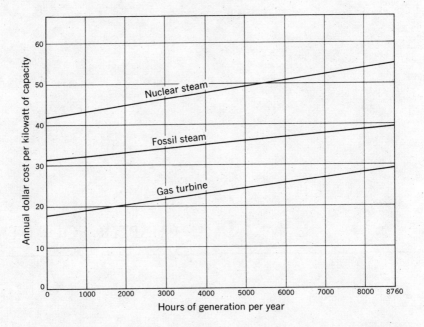

Figure 13.5. Hypothetical screening curves for the generation of electricity assuming unrestricted imports of cheap oil. Nuclear and high-performance fossil-fuel steam technologies would not be competitive.

very hypothetical example, we see that international oil competition and collapse of the United States oil price would destroy not only the nation's fossil-fuel industry infrastructure, but the nuclear and high-performance fossil-fuel electric power generation infrastructures as well, with additional loss of jobs and investments. This potentiality provides additional support for my basic assumption that the United States fossil fuel industries will continue to be protected.

CHAPTER FOURTEEN

COMPARISON OF FOSSIL FUEL,

NUCLEAR, AND SOLAR ELECTRIC

ECONOMIES

Fossil fuels, nuclear fuels, and solar energy are almost certain to continue as the major energy sources of the future, and it is the purpose of this chapter to estimate their future competitiveness in the light of probable technological progress.

Noneconomic factors are of fundamental significance to the future of the energy industries. Every forecast of fuel availability, energy consumption, conversion technologies, and electrification must be based implicitly or explicitly on an assessment of political and social forces that provide the overall environment—the basic ground rules—within which economics and technology operate. We have noted that unrestricted international competition in petroleum would drop the price of crude oil five-fold or more in the United States, ruining the nation's oil and coal industry infrastructures and ruining the world's nuclear and high-performance steam turbine infrastructures. In such a world, economics and technology would lead to the simple gas turbine as prime mover for electric power generation. We have also noted that unrestricted international competition seems highly unlikely in view of the nearly universal opposition to it by most interested parties, and that protection of energy industry infrastructures is more probable. In such a world, economics and technology would lead to further economies of scale and of new

131

technology in the recovery and refining of fossil fuels, including the production of synthetic crude oil from solid fuels, to further development of nuclear energy, and to more efficient and lower cost conversion equipment. This widespread technological activity stands in sharp contrast to the decay and stagnation that would accompany a collapse in the price of oil. Clearly, it makes no sense at all to discuss economics and technology without first assessing the probable future political and social environment.

Three fundamental political factors were identified in Chapter 7 and are restated here: the oil-producing and exporting countries want to limit production, keep prices from collapse, and maximize their total rent from the United States; the United States wants to protect its energy industry infrastructure for a variety of reasons, including preservation of jobs and investments; and consumers are (or in the event would be) fearful of exposure to monopolistic powers beyond the control of their government. These factors strongly suggest for the United States that the energy industry infrastructure will be preserved, that fossil fuel production will not decline, and that prices will continue to be determined by the costs of domestic marginal production. It may be that the job will be accomplished by tariffs and quotas, but by whatever method the infrastructure is likely to be preserved.

Social factors are also of significance in providing ground rules for the operation of economics and technology, and the environmental movement is primary among them. The combustion of unrefined fuel is in effect being outlawed by restrictions on air pollution, and restrictions on the use of cooling water are causing a shift to air cooling. The higher cost of refined fuel (or equivalently of stack-gas cleanup) and the higher cost of air cooling constitute new motivations for technology to improve the efficiency of electric generation.

The two noneconomic factors just reviewed, one political and one social, in large part determine the overall environment in which the technological and economic struggles for position among fuels and energy conversion technologies will be fought out in the United States. The political factor, in conjunction with vast fuel resources, suggests that fossil fuels will continue to be available at historic prices, in adequate amounts to support total anticipated energy consumption for many centuries, and that nuclear fuels will be available for many millenia. The social factor suggests that the social costs of energy-related activities will be internalized. Although the future is uncertain, the political and social environment just described provides one view of what the future may be like, and I use it as the basis for the rest of this chapter.

Because the anticipated political and social environment amounts largely to a preservation of past self-sufficiency in energy supplies within the United States, the projections of energy consumption made on that basis in Chapter 3 can be considered more seriously. Per capita consumption of energy is projected to continue to grow for several more decades, finally levelling off at about 1 1/2 times current consumption early in the next century. Combined with population growth, this gives an expanding market, albeit with a declining growth rate, for further development and evolution of the energy industries in the United States. Since the political model envisions the internal energy economics and technology of the United States as being largely disconnected from the rest of the world, it may happen that other parts of the world will follow different paths, with different fuel prices, different technologies, and different levels of energy consumption. Because political and social factors are overriding in these matters, each region, such as Europe, Japan, or the Soviet Union, requires individual analysis. Hence the growth in energy consumption embodied in Figure 3.2 should be considered as applying to the United States only, not necessarily to other industrialized regions.

Fuel price levels within the United States are projected to continue their historic decline (in constant dollars) as analyzed in Chapters 5 and 8.

For electric utility generation (but not for transportation or space heating, which already use relatively expensive clean fuel), the cost of all forms of fossil fuel is likely to double. Natural gas price at the wellhead—near where most gas-burning electric utilities get it—may approximately double when natural gas finds its competitive level. (Owing to relatively large transportation and distribution costs, doubling the wellhead price in the Southwest increases the price to a residential consumer in the Northeast by only about 20 percent.) The cost of utility coal may approximately double, because of increased transportation required to bring low-sulfur coal from long distances or because of the necessity for refining coal to purify it. The cost of utility oil may approximately double as utilities shift from high-sulfur residual oil (the dregs of refining) to the more expensive low-sulfur oil used by others.

The approximate doubling of fossil-fuel prices is a competitive boon to nuclear power, for it increases the cost of electricity from fossil fuel relative to electricity from nuclear fuel. There is a tendency among nuclear power advocates to hope that one doubling will be followed by others, and to project much higher fossil fuel prices in the future; but I believe this view mistaken. Social and environmental factors are requiring the utility industry to join the rest of the country in burning clean fuel purchased at competitive prices. The price of clean fuel itself is likely to continue its historic downward trend.

When measured in constant dollars per kilowatt of capacity, the cost of constructing a nuclear power plant increased by perhaps 50 percent in the past decade during which time the cost of a fossil-fuel steam plant increased slightly and the cost of a gas turbine power plant declined. When power plant costs rise an explanation is required, as we expect all power plant costs to decline through the economies of scale and new technology. The environmental movement was responsible for part of the rise in nuclear plant costs, by causing various procedural delays and by requiring additional expensive safeguards to protect against hypothethical accidents. But there appears to be another cause for increasing construction costs, associated with a growing proportion of high-cost field construction and a shrinking proportion of low-cost factory construction for the very large power plants now being built. As sketched in Appendix 5, the costs associated with a shift to field from factory can more than offset anticipated economies of scale. Gas turbine power plants have little field construction labor content and hence respond as anticipated to economies of scale and new technology. Fossil-fuel steam plants, following nuclear plants toward larger sizes and an increasing proportion of field construction (but on average lagging by a factor of 2 in size) represent an intermediate case.

Whatever its causes, the increase in nuclear power plant construction costs has dealt a competitive blow to nuclear power, in effect cancelling the competitive advantage conferred by doubled fossil fuel prices. Advocates of fossil fuel may tend to hope that one 50 percent increase in nuclear plant construction costs will be followed by others and to project much higher nuclear plant costs in the future; but I believe this view mistaken. The field construction cost lesson appears to have been learned, and there is an increasing trend toward standardization and assembly line construction of major nuclear power plant components. The requirements of the new licensing procedures are being better anticipated now that they are more familiar, and licensing delays may diminish. Overall, competition between nuclear power and fossil-fuel power is likely to remain vigorous.

Turning now to the pace and extent of future electrification, we note that both fossil-fuel and nuclear-fuel conversion technologies are under increased pressure to improve efficiency. Fossil-fuel conversion efficiency is more important than before because the refined fuel required today is more costly than the formerly acceptable unrefined fuel. Nuclear-fuel conversion efficiency is more important than before because construction costs are higher, and cost per unit electrical output can be decreased by raising the output from a plant of a given size. Both technologies are moving ahead rapidly, and from the overall standpoint of electrification, one or the other or both in combination are expected to enable electrification to progress for decades, perhaps at a pace close to the intermediate projection in Figure 11.6. Interest is intense in the future market shares that will be achieved by nuclear and fossil fuels and technologies, but prudence suggests that a detailed forecast not be made. I am content to forecast continued electrification and to watch with interest the competitive battle among the various fuels and technologies.

Although solar energy is currently without commercially active champions, except for the builders of hydroelectric power plants who are not on an upward trajectory, it is important to include solar energy on the same basis as fossil and nuclear fuels to make the record complete. Except for hydropower, generation of steam from the combustion of agricultural products appears to be the most economic path for solar energy utilization, and this technology will be compared with fossil and nuclear technologies. Agricultural products have the virtue of collecting both direct solar radiation such as can be concentrated by a lens and diffuse radiation such as is received on cloudy days. The energy is stored in chemical form for later combustion when needed. The products of combustion are nonpolluting and are recycled through the atmosphere. Of all the possible agricultural products that might be considered, grass is one of the most attracive. In the form of hay, it is a commodity with United States production of 127 million tons valued at $3.1 billion in 1970 (14.1). Its price is well established and its potential future price can be estimated.

The previous chapter contained a rough comparison for nuclear steam, fossil-fuel steam, and gas turbine technologies in the early 1970's, considering only power plant fixed charges and fuel costs. Had transmission fixed charges and transmission losses been considered, they would have added to the cost of electricity generated by steam plants which require transmission, but would not necessarily have added to the cost of electricity generated by smaller gas turbine plants, which often are sited at substations closer to the load. Had operation and maintenance costs been included, they would have added more to the cost of electricity generated by gas turbines, tending to restore the balance to that shown in Figure 13.4. Hence overall the picture in Figure 13.4 is a fair approximation of the relative competitiveness of fossil and nuclear fuel technologies in the early 1970's. The data on which Figure 13.4 is based are repeated in Table 14.1, with data for solar steam technology added. The solar steam technology is simply a duplicate of fossil-fuel steam technology

TABLE 14.1 ILLUSTRATIVE COST ELEMENTS FOR ELECTRIC POWER GENERATION IN EARLY 1970'S

Type of Plant	Capital Cost	Heat Rate	Fuel Cost
Nuclear steam	$280/kW	10,000 Btu/kWh	15¢/million Btu
Fossil steam	$210/kW	9,000 Btu/kWh	40¢/million Btu
Gas turbine	$120/kW	13,000 Btu/kWh	70¢/million Btu
Solar steam	$210/kW	9,000 Btu/kWh	200¢/million Btu

with hay at $2 per million Btu (the going price for hay) substituted for fossil fuel at 40¢ per million Btu. With a five-fold increase in fuel cost over a fossil steam plant, it is easy to see why the hay-burning steam plant is not commercially attractive.

Looking ahead to the end of the century, we may anticipate changes in the cost elements for various generation technologies, and we may speculate on their magnitude. For definiteness and convenience, all cost elements postulated for the year 2000 will be expressed in 1970 dollars.

Considering first nuclear steam, we note that the capital cost of plants under construction today and scheduled for commercial operation in 1980 is about $280 per kilowatt. Assuming that the environmental movement holds no more hidden surprises, and that nuclear power continues to grow as electrification increases, economies of scale and of new technology could reduce this cost to perhaps $200 per kilowatt by 2000. Nuclear fuel cost has been driven up by social factors to where it is closer to 20¢ per million Btu for the 1980 time frame than to the 15¢ per million Btu of the early 1970's. The economy of scale may bring it back to 15¢ per million Btu by 2000. The nuclear steam heat rate may be improved to perhaps 8000 Btu per kilowatt-hour.

The capital cost of fossil-fuel steam plants could be reduced to perhaps $140 per kilowatt. And through new technology such as a very-high-temperature alloy steel boiler or more likely a gas turbine and steam turbine combined cycle, a heat rate of 6500 Btu per kilowatt-hour is achievable. Fuel, which cost 40¢ per million Btu for an oil-fired fossil-fuel steam plant before the Clean Air Act was implemented, is likely to cost more in 2000 when full implementation is the rule, although somewhat less than clean fuel would cost today because of the additional cost-reducing experience to be built up in the fuel industries. A fuel cost of 60¢ per million Btu might be appropriate.

The cost of energy from hay depends on whether new strains of grass are bred for their heating value. If we imagine that agricultural scientists have been active in breeding hay with low nitrogen content and high cellulose content, designed to store solar energy for combustion instead of for feed, a fuel hay cost of 60¢ per million Btu may be achievable (14.2). This would make solar steam an even competitor with fossil-fuel steam. Equality of the costs of fossil fuel and hay, if ever achieved, would not mean that hay would take over from fossil fuels. At most, it would allow hay to become an alternate combustible fuel and the marginal combustible fuel, so that the price of hay would subsequently limit United States fossil fuel prices. Although I believe it is possible, I do not expect that it will happen. Should it happen, it could only gradually affect the nation's energy economics.

The hypothetical cost elements developed for the year 2000 are shown in Table 14.2. Corresponding screening curves showing annual cost of electric

TABLE 14.2   HYPOTHETICAL COST ELEMENTS FOR ELECTRIC POWER
GENERATION IN 2000 (1970 DOLLARS)

Type of Plant	Capital Cost	Heat Rate	Fuel Cost
Nuclear steam	$200/kW	8,000 Btu/kWh	15¢/million Btu
Fossil steam	$140/kW	6,500 Btu/kWh	60¢/million Btu
Solar steam	$140/kW	6,500 Btu/kWh	60¢/million Btu

generation as it depends on annual hours of operation are plotted in Figure
14.1, which may be compared with Figure 13.4 for the early 1970's. The
hypothetical cost elements for the year 2000 were chosen in part to show the
potential future competitiveness of nuclear, fossil-fuel, and solar steam
technologies. Other technologies, including in particular fuel cell generation,
also have potential for competitiveness.

Although it is not possible from this vantage point to determine what combination of fuels and technologies will be leaders in the next century, we are as

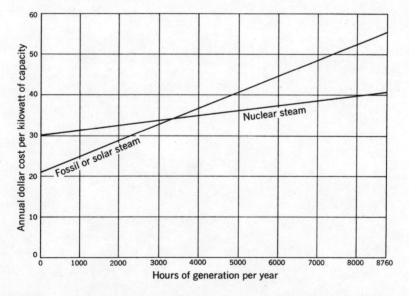

Figure 14.1. Screening curves for potential nuclear, fossil-fuel, and solar steam
technologies of 2000. These curves are not projections or forecasts. They are intended
only to suggest the potential future competitiveness of technologies based on nuclear,
fossil-fuel, and solar-energy resources.

fortunate with technologies as with fuels in having many viable alternatives. We possess the technologies that will enable the cost of electricity to be reduced, the efficiency of generation to be increased, and electrification to proceed.

CHAPTER FIFTEEN

ENERGY CRISES

IN PERSPECTIVE

The energy industries are managed by men whose decisions are based on anticipated future conditions. If their anticipations are confounded by political or social upheavals, of which the Suez Canal closing and the environmental movement are examples, an energy crisis can result. The closing of the Suez Canal in 1967 precipitated a crisis in Europe because a larger tanker fleet was suddenly and unexpectedly needed to carry oil around Africa to Europe. Within a few years, the tanker fleet was built and the crisis was over. The environmental mass movement of the late 1960's and the early 1970's precipitated a crisis in the United States because more petroleum products and more refinery capacity were suddenly and unexpectedly needed as consumers shifted to oil from coal and from planned uranium. Within a few years, the refinery capacity will be built and the petroleum products will be flowing.

Failure to act in the face of uncertainty about future conditions also can lead to crisis. The uncertain viability of high-cost crude oil and refined products in the face of a potential flood of low-cost imported crude oil and refined products inhibited United States petroleum development and refinery construction, adding to the current crisis. Such uncertainty can only be dispelled, and action initiated, by a clear government policy for protecting energy industry investments and infrastructure.

Government policy can cause crisis as well as alleviate it. When regulated fuel prices are held so low that resource development is inhibited while demand is inflated, supply and demand can move so far apart that shortages and rationing result, as has happened with natural gas.

Other crises of similar nature are sure to occur in years to come. Future conditions will be wrongly anticipated again, putting some segments of the energy industry into inappropriate positions. Future conditions will be clouded again, inhibiting action until the situation is clarified. Future regulations may interfere with the marketplace again, upsetting supply and demand once more. But all crises of this nature tend to be of limited duration and self-correcting. As events reveal mistakes of anticipation, the mistakes are corrected. Decisions delayed are finally made. Counterproductive regulations are finally abandoned. Perhaps the greatest danger to the United States energy industry infrastructure, as indeed to the energy industry infrastructure of every industrialized nation, is its potential destruction by a flood of low-cost foreign oil imports, a danger that I assume is so well recognized and so politically potent to workers, investors, and consumers that government policy is sure to prevent it.

Although crises may be expected from time to time, the fundamentals of the energy industries are sound. Fossil fuels, nuclear fuels, and solar energy resources are abundant. Improved technologies for generating and utilizing electricity can increase the extent of electrification. The greatest imponderables may be the future acceptability and price of the various forms of energy— acceptability as affected by ecological and environmental pressures and price as affected by domestic and international political pressures.

Although no one can be sure what the future will bring, I would suggest, as worthy of the reader's consideration, a potential future for the United States in which environmental pressures require that only refined fuels be burned, and that lands disturbed by mining be reclaimed; and in which political pressures require that oil imports be limited so that domestic production does not shrink, with the consequence that existing marginal producers continue to determine domestic fuel prices. Under these conditions the United States energy economy, with its moderate fuel prices and high technology content, might be expected to continue its historic evolution. Through further progress of technology and further economies of scale, particularly with the advent of synthetic crude oil from coal or shale, prices of refined fuels (and of electricity after adjusting for the shift to refined fuel) might be expected to continue their downward trends, when measured in constant dollars to strip away the effects of inflation.

The path to the future will not be easy. Development and implementation of new technologies for synthetic fuel, nuclear power, and electric energy conversion will require the sustained efforts of many dedicated people. Yet even though some attempts may fail, I am confident that others will succeed, and that the nation's energy economy can continue its historic evolution.

REFERENCES

1.1.   In Figure 1.1, data for consumption of fossil fuel were taken from *United Nations Statistical Papers*, Series J, No. 15, "World Energy Supplies 1961–1970," United Nations, New York, 1972. Data for numbers of work animals and for per capita consumption of agricultural wastes, dung, and fuel wood were taken from Robert A. Harper, "The Geography of World Energy Consumption," *The Journal of Geography*, October 1966, pages 302–315. Per capita food energy was estimated to be approximately 2000 calories daily in the nonindustrialized regions and 3000 calories daily in the industrialized regions. Feed energy was taken as 90 million Btu per animal per year in industrialized regions, this being the gross energy content of work animal feed as estimated in Appendix 2. In the nonindustrialized regions a work animal is fed an estimated two-thirds as much, or about 60 million Btu every year, of which about half is recovered in the form of dung and is utilized as fuel. To avoid double counting, only 30 million Btu of the 60 million Btu input can be counted as (net) feed energy, since 30 million Btu is being counted as dung.

1.2.   Stanford Research Institute, for the Office of Science and Technology, Executive Office of the President, Washington, D. C., *Patterns of Energy Consumption in the United States*, U.S. Government Printing Office, Washington, D.C., January 1972.

1.3.   U.S. Bureau of the Census, *Statistical Abstract of the United States: 1972.* (93d edition.) Washington, D.C., 1972.

1.4.   The Stanford Research Institute study (1.2) did not give data for lighting as a separate end use in 1960. Hence in determining growth rates it was not possible to treat lighting as a basic energy use where it most properly belongs. Lighting had to be included with new energy uses where it was classified in 1960.

2.1.   *The World Almanac*, 1972 edition, Newspaper Enterprise Association, New York.

2.2.   *United Nations Department of Economic and Social Affairs, Technical Papers*, Series J, No. 15, "World Energy Supplies 1961–1970," United Nations, New York, 1972.

4.1.   Estimates by M. King Hubbert in National Academy of Sciences—National Research Council, *Resources and Man*, W. H. Freeman, San Francisco, 1969.

4.2.   P. K. Theobald, S. P. Schweinfurth, and D. C. Duncan, *Energy Resources of the United States*, Geological Survey Circular 650, U.S. Geological Survey, Washington, D.C., 1972.

4.3.   A. B. Cambel, et al., *Energy R & D and National Progress: Findings and Conclusions*. U.S. Government Printing Office, Washington, D.C., 1966.

4.4.   Nonenergy uses of fossil fuels in the United States have grown rapidly in the past half century to where they consume annually about 1000 pounds of fossil fuel per capita. Most of these uses amount to the substitution of a fossil-fuel product for a wood product. Asphalt and road oil can be viewed as replacing planks and logs in highway construction. Asphalt tiles and shingles can be viewed as replacing wooden flooring and shingles. Plastics of various types can be viewed as replacing wood in all of its other applications. Although synthetic fibers are replacing natural fibers such as silk, cotton, wool, and linen, the quantities involved are relatively small. Wood is the main target for displacement.

In the 70-year period from 1900 to 1970, the combined per capita consumption of roundwood (wood as cut before being made into wood products) and of fossil fuels for nonenergy uses averaged about 2300 pounds annually, with fluctuations reaching a high of nearly 3000 pounds in 1905 and a low near 1500 pounds in 1935. Although the combined per capita consumption has been essentially constant, fossil fuel's share has risen steadily to 38 percent in 1970 and is projected to reach 50 percent in the mid-1980's. I assume that the ultimate nonenergy consumption of fossil fuels will saturate at about 2300 pounds annually per capita when wood is completely displaced. This corresponds roughly to 50 million Btu of fossil fuels annually per capita.

4.5.   Most of the heat that flows outward from deep inside the earth flows very slowly by thermal conduction through solid rock. (Heat carried along by the direct flow of molten rock, as from erupting volcanos, is a minor exception.) If the geothermal heat—thermal energy slowly on its way up to the surface—should be removed by man from the uppermost few miles of rock, many millions of years would be required for it to be replenished by conduction from below. Hence geothermal energy may properly be considered a nonrenewable resource on any time scale of significance to mankind.

5.1.   G. Manners, *The Geography of Energy*, Hutchinson, London, 1971.

5.2.   *Bureau of Mines Information Circular 8535*, "Cost Analyses of Model Mines for Strip Mining of Coal in the United States," U.S. Department of the Interior, Bureau of Mines, 1972.

5.3.   Since about two barrels of crude oil are consumed in producing one barrel of gasoline and one barrel of other products, a question arises as to what fraction of the crude oil cost should be charged against gasoline and what fraction against the other products. If the barrel of other products were worthless, the barrel of gasoline should be charged with the cost of two barrels of crude oil. If the other products were as valuable as gasoline, each should be charged with the cost of one barrel of crude oil. The situation is intermediate, with gasoline somewhat more valuable than the other products, and I estimate that the barrel of gasoline should be charged with the cost of 1.2 barrels of crude oil. The gasoline processing costs plotted in Figure 5.1 are retail gasoline prices (exclusive of tax) per barrel, less 1.2 times the wellhead price of crude oil per barrel (5.4). All prices are measured in 1970 dollars and are assumed to approximate marginal cost (including a minimal return). Cumulative gasoline production was computed from data in Reference 5.4.

5.4.   *Petroleum Facts & Figures, 1971 Edition.* American Petroleum Institute, Washington, D. C.

5.5.   H. C. Hottel and J. B. Howard, *New Energy Technology—Some Facts and Assessments,* MIT Press, Cambridge, Mass., 1971.

5.6.   Hans Rumpf, *The Bombing of Germany,* translated by Edward Fitzgerald, Holt, Rinehart and Winston, New York, 1963.

6.1.   Electric Power Survey Committee of the Edison Electric Institute, *Semi-annual Electric Power Survey* as of April 1 and October 1 every year; also the *Year-end Summary of the Electric Power Situation in the United States* as of December 31 every year; The Edison Electric Institute, New York.

6.2.   Edison Electric Institute, *Statistical Year Book of the Electric Utility Industry* for various years, Edison Electric Institute, New York; also *Historical Statistics of the Electric Utility Industry,* Edison Electric Institute, New York, 1969.

7.1.   M. A. Adelman, *The World Petroleum Market,* published for Resources for the Future, Inc. by The Johns Hopkins University Press, Baltimore, 1972.

11.1.   N. B. Guyol, *The World Electric Power Industry,* University of California Press, Berkeley, 1969 for the years 1920, 1929, 1937, 1950. Federal Power Commission, *World Power Data 1969,* U.S. Government Printing Office, Washington, D.C., 1972 for 1955–1969.

11.2.   See Reference 6.2.

11.3.   Data for the energy input to U.S. electric power generation (including the fuel equivalent of hydroelectric power) were approximated from Reference 6.2 by applying the electric utility heat rate to the total United States electric power generation. Data for total energy input to the United States were taken from Table 3, Appendix 2.

11.4.   Federal Power Commission, *The 1970 National Power Survey, Federal Power Commission, Part I,* Washington, D.C., 1971; *Statistics of Privately Owned Electric Utilities in the United States 1970,* Washington, D.C., 1971; *Statistics of Publicly Owned Electric Utilities in the United States 1970,* U.S. Government Printing Office, Washington, D.C., 1972.

11.5.   See Reference 1.2. The Stanford Research Institute analysis (1.2) and the Edison Electric Institute analysis (6.2) disagree slightly but not significantly with respect to the improvement in efficiency of electric power generation between 1960 and 1968. I use the Stanford Research Institute data for my analysis of electrification in Chapter 11.

11.6.   Data from Appendix 1 based on Reference 1.2.

11.7.   Projections of energy consumption for all end uses, for end uses except process steam and transportation, and for the lower bound on energy utilized via electricity are based on per capita consumption projections developed in Appendix 3 combined with an annual population growth rate of 1.3 percent. Except for its end points, the location of the upper bound is more subjective, and it was simply drawn in freehand.

14.1.   See Reference 1.3.

14.2.   See William J. Oswald and Clarence G. Golueke, "Biological Transformation of Solar Energy," a chapter in *Advances in Applied Microbiology, Volume 2,* edited by Wayne W. Umbreit, Academic Press, New York, 1960. The authors discuss the conversion of light energy into chemical energy by algae growing in ponds, and report efficiencies ranging from 2 to 10 percent in various experiments. I assume that agricultural scientists would be able to breed strains of grass with similar conversion efficiency, and with low nitrogen content and high cellulose content designed to store chemical energy for combustion. Today's average hay crop of about 2 tons per acre per year collects solar energy with something like 0.15 percent efficiency. Assuming that the efficiency could be increased sixteen-fold to 2.4 percent with new strains of grass and special farming techniques, and assuming that it might cost the farmer 4 times as much per acre to grow the new hay, the cost of hay energy could be reduced to about 60¢ per million Btu as delivered to a nearby power plant.

APPENDIX 1

PATTERNS OF ENERGY CONSUMPTION

IN THE UNITED STATES

The data summarized in this appendix are taken from the Stanford Research Institute (SRI) report *Patterns of Energy Consumption in the United States,* Office of Science and Technology, Executive Office of the President, Washington, D.C., January 1972.

The SRI report measures hydroelectricity by its direct energy content. However, as discussed in Chapter 2 of this book, it is possible alternatively to measure hydroelectricity by the larger amount of energy that would be required to generate the same amount of electricity in a fuel-burning power plant. Because I prefer this alternate way of measuring hydropower, I have converted the SRI energy values for hydropower to their fuel-input equivalents. This causes small changes in many of the SRI numbers, making it desirable to list the amended values here.

Table A1.1 lists in the first column the amended fuel energy inputs for various end uses in the United States in 1968 from the SRI report, makes adjustments based upon additional information given in the report, and finally strips out "feedstocks," which are fossil fuels used for nonenergy purposes to obtain the last column with percentages of energy utilized for various end uses.

Table A1.2 summarizes growth rates of the various end uses for energy between 1960 and 1968, according to the three end use groupings: basic energy uses in commerce, industry and transportation; basic residential energy uses; and new energy uses.

TABLE A1.1 SIGNIFICANT END USES FOR ENERGY (U.S., 1968)[a]

End Use	Fuel[a] (trillions of Btu)	Fuel (%)	Fuel Adjustments (%)	Fuel (revised %)	Energy (%)
Transportation[b]	15,044	24.2		24.2	25.6
Space heating	10,980	17.7	+2.0[g]	19.7	20.8
Process steam	9,271	14.9	−1.1[g]	13.8	14.6
Direct heat	7,007	11.3	−0.9[g]	10.4	11.0
Industrial drive	6,068	9.8		9.8	10.3
Feedstocks	3,332	5.4		5.4	—
Water heating	2,496	4.0		4.0	4.2
Refrigeration[c]	1,784	2.9		2.9	3.0
Lighting	f	f	+2.7[h]	2.7	2.8
Air conditioning[d]	888	1.4		1.4	1.5
Electrolytic processes	882	1.4		1.4	1.5
Cooking	812	1.3		1.3	1.4
Television	f	f	+0.6[i]	0.6	0.7
Clothes drying	227	0.4		0.4	0.4
Other (electric)[e]	3,325	5.3	−3.3[h,i]	2.0	2.2
	62,116	100.0		100.0	100.0

[a] *Patterns of Energy Consumption in the United States*, Office of Science and Technology, Executive Office of the President, Washington, D.C., January 1972. From Figure 2, amended to include hydroelectric energy at fuel equivalent, and corrected for an error in commercial air conditioning.

[b] Includes farm work, other nonfactory work, automobile air conditioning.

[c] Includes home freezers.

[d] Does not include automobile air conditioning.

[e] Includes small appliances, computers, elevators, escalators, office machinery, commercial electric heat, industrial air conditioning, and other miscellaneous end uses.

[f] Included in "other".

[g] Footnote a, page 8, estimates that industrial space heating, included in process steam and direct heat in column 1, amounts to about 2 percent of total United States energy consumption.

[h] Footnote a, page 62 for residential sector. For commercial sector, lighting was estimated to be 55 percent of "other" category (after correction of error in air conditioning). For industrial sector, lighting was estimated to be half of "other" category.

[i] Footnote a, page 12.

TABLE A1.2   ENERGY USE GROWTH RATES

Energy Uses	Energy (in trillions of Btu)		% Annual Growth
	1960	1968	
Basic energy uses in commerce, industry, and transportation			
Transportation	10,882	15,044	4.1
Space heating[a]	3,126	4,248	3.9
Process steam	6,858	9,271	3.8
Direct heat	5,616	7,007	2.8
Industrial drive	4,296	6,068	4.4
Water heating	578	682	2.1
Electrolytic processes	660	882	3.7
Cooking	100	141	4.4
	32,116	43,343	3.8
Basic residential energy uses			
Space heating	4,861	6,732	4.2
Water heating	1,231	1,814	5.0
Cooking	592	671	1.6
	6,684	9,217	4.1
New energy uses			
Refrigeration	1,151	1,784	5.6
Air conditioning	376	888	11.3
Television	191	398	9.6
Clothes drying	104	227	10.2
Other electric	1,428	2,927	9.4
	3,250	6,224	8.5
All energy uses	42,050	58,784	4.3

[a] Industrial space heating included in process steam and direct heat.

TABLE A1.3   SIGNIFICANT END USES FOR
ELECTRICITY (U.S., 1968)[a]

End Use	Electric Energy (fuel equivalent, trillions of Btu)	Electric Energy (%)
Industrial drive	6,068	39.7
Refrigeration[b]	1,779	11.6
Lighting	1,649	10.8
Water heating	949	6.2
Electrolytic processes	882	5.8
Air conditioning	854	5.6
Space heating	508	3.3
Direct heat	403	2.6
Television	398	2.6
Cooking	321	2.1
Clothes drying	160	1.0
Transportation	55	0.4
Other[c]	1,278	8.3
	15,304	100.0

[a] *Patterns of Energy Consumption in the United States*, Office of Science and Technology, Executive Office of the President, Washington, D.C., January 1972. Amended to include hydroelectric energy at fuel equivalent, and corrected for an error in commercial air conditioning.
[b] Includes home freezers.
[c] Includes small appliances, computers, elevators, escalators, office machinery, commercial electric heat, industrial air conditioning, and other miscellaneous end uses.

Table A1.3 lists the significant end uses for electricity, first by the fuel input to generation (including the fuel equivalent of hydroelectricity), then by percent.

Table 9.1 in Chapter 9 separates the end uses of electricity into three groupings: electrified, partially electrified, and unelectrified. Table 11.2 in Chapter 11 summarizes the growth rates of the various electrified end uses between 1960 and 1968 according to the same groupings.

APPENDIX 2

FUEL AND ENERGY INPUTS TO

THE UNITED STATES 1850 TO 1970

A number of different sources provided significant energy inputs to the United States at one time or another in the 120 years from 1850 to 1970. In order of their historical development, they can be classified under the headings of solar energy, fossil fuels, and nuclear fuels:

SOLAR ENERGY: conversion via
    Fuel wood
    Work animal feed
    Direct windpower and waterpower
    Hydroelectricity

FOSSIL FUELS: combustion of
    Anthracite coal
    Bituminous coal and lignite
    Petroleum
    Natural gas

NUCLEAR FUELS: fission of
    Uranium and thorium.

Table A2.1 summarizes the annual fuel input to the United States for the years 1850–1970, including nonenergy uses of the fossil fuels which in recent years have amounted to about 1 percent of coal consumption and about 7 percent of oil and gas consumption. The numerical values for each fossil fuel, fuel wood, and animal feed are the amounts of heat they would generate

TABLE A2.1 CONSUMPTION OF FOSSIL FUELS, SOLAR ENERGY, AND NUCLEAR ENERGY: THE UNITED STATES, 1850–1970

[in units of 1C = 10^{16} Btu]

		Fossil Fuels			Solar Energy						Nuclear Energy
	Total Fossil Fuels	Coal	Petroleum	Natural Gas (dry)	Total Solar Energy	Fuel Wood	Work Animal Feed	Direct Wind and Water Power (feed equivalent)	Hydroelectric Power (fuel equivalent)	Total Nuclear Energy
Year	Total Consumption										
	1	2	3	4	5	6	7	8	9	10	11
1970	6.8472	6.4565	1.2922	2.9614	2.2029	0.3678	0.0900	0.0108	0.0020	0.2650	0.0229
1969	6.6038	6.2175	1.2733	2.8422	2.1020	0.3717	0.0910	0.0126	0.0022	0.2659	0.0146
1968	6.2852	5.9291	1.2659	2.7052	1.9580	0.3431	0.0920	0.0145	0.0024	0.2342	0.0130
1967	5.9383	5.5841	1.2256	2.5335	1.8250	0.3462	0.0930	0.0162	0.0026	0.2344	0.0080
1966	5.7560	5.4282	1.2495	2.4395	1.7392	0.3221	0.0940	0.0180	0.0028	0.2073	0.0057
1965	5.4521	5.1247	1.1908	2.3242	1.6097	0.3236	0.0950	0.0198	0.0030	0.2058	0.0038
1964	5.2445	4.9299	1.1264	2.2387	1.5648	0.3111	0.0960	0.0212	0.0032	0.1907	0.0035
1963	5.0539	4.7507	1.0714	2.1950	1.4843	0.2998	0.0970	0.0227	0.0034	0.1767	0.0034
1962	4.8679	4.5577	1.0189	2.1267	1.4121	0.3078	0.0980	0.0241	0.0036	0.1821	0.0024
1961	4.6603	4.3621	0.9906	2.0487	1.3228	0.2964	0.0990	0.0256	0.0038	0.1680	0.0018
1960	4.5887	4.2906	1.0140	2.0067	1.2699	0.2975	0.1000	0.0278	0.0040	0.1657	0.0006
1959	4.4483	4.1548	0.9810	1.9747	1.1991	0.2933	0.1013	0.0287	0.0042	0.1591	0.0002
1958	4.3075	4.0059	0.9849	1.9214	1.0996	0.3014	0.1026	0.0308	0.0044	0.1636	0.0002

Year											
1957	4.3118	4.0154	1.1168	1.8570	1.0416	0.2963	0.1039	0.0327	0.0046	0.1551	0.0001
1956	4.3156	4.0212	1.1752	1.8625	0.9835	0.2944	0.1053	0.0356	0.0048	0.1487	—
1955	4.1208	3.8296	1.1540	1.7524	0.9232	0.2912	0.1067	0.0388	0.0050	0.1407	—
1954	3.7832	3.4875	1.0195	1.6132	0.8548	0.2957	0.1086	0.0431	0.0052	0.1388	—
1953	3.9237	3.6149	1.1893	1.6099	0.8157	0.3088	0.1105	0.0490	0.0054	0.1439	—
1952	3.8199	3.4962	1.1868	1.5334	0.7760	0.3237	0.1124	0.0562	0.0055	0.1496	—
1951	3.8623	3.5321	1.3225	1.4848	0.7248	0.3302	0.1144	0.0647	0.0057	0.1454	—
1950	3.5932	3.2551	1.2913	1.3488	0.6150	0.3381	0.1164	0.0718	0.0059	0.1440	—
1945	3.3832	3.0055	1.5972	1.0110	0.3973	0.3777	0.1261	0.1098	0.0081	0.1337	—
1940	2.6643	2.3042	1.2535	0.7781	0.2726	0.3601	0.1358	0.1330	0.0088	0.0825	—
1935	2.2089	1.8288	1.0634	0.5680	0.1974	0.3801	0.1397	0.1530	0.0126	0.0748	—
1930	2.5580	2.1506	1.3639	0.5898	0.1969	0.4074	0.1455	0.1750	0.0163	0.0706	—
1925	2.4774	2.0198	1.4706	0.4280	0.1212	0.4576	0.1533	0.2142	0.0270	0.0631	—
1920	2.4201	1.9007	1.5504	0.2676	0.0827	0.5194	0.1610	0.2508	0.0378	0.0698	—
1915	2.0670	1.5385	1.3294	0.1418	0.0673	0.5285	0.1688	0.2635	0.0340	0.0622	—
1910	1.9301	1.4261	1.2714	0.1007	0.0540	0.5040	0.1765	0.2489	0.0301	0.0485	—
1905	1.5783	1.0983	1.0001	0.0610	0.0372	0.4800	0.1843	0.2318	0.0292	0.0347	—
1900	1.2086	0.7322	0.6841	0.0229	0.0252	0.4764	0.2015	0.2241	0.0283	0.0225	—
1895	1.0156	0.5265	0.4950	0.0168	0.0147	0.4891	0.2306	0.2215	0.0289	0.0081	—
1890	0.9307	0.4475	0.4062	0.0156	0.0257	0.4832	0.2515	0.2002	0.0295	0.0020	—
1885	0.7628	0.2962	0.2840	0.0040	0.0082	0.4666	0.2683	0.1677	0.0306	—	—
1880	0.6780	0.2150	0.2054	0.0096	—	0.4630	0.2851	0.1463	0.0316	—	—
1875	0.5924	0.1451	0.1440	0.0011	—	0.4473	0.2872	0.1291	0.0310	—	—
1870	0.5365	0.1059	0.1048	0.0011	—	0.4306	0.2893	0.1109	0.0304	—	—
1865	0.4811	0.0642	0.0632	0.0010	—	0.4169	0.2767	0.1086	0.0316	—	—
1860	0.4554	0.0521	0.0518	0.0003	—	0.4033	0.2641	0.1064	0.0328	—	—
1855	0.3992	0.0421	0.0421	—	—	0.3571	0.2389	0.0899	0.0283	—	—
1850	0.3329	0.0219	0.0219	—	—	0.3110	0.2138	0.0734	0.0238	—	—

TABLE A2.2 CONSUMPTION OF FOSSIL FUELS, SOLAR ENERGY, AND NUCLEAR ENERGY: THE UNITED STATES, 1850–1970

[in % of total consumption]

| Year | Total Consumption | Fossil Fuels | | | | Solar Energy | | | | Hydro-electric Power (fuel equivalent) | Nuclear Energy |
| | | Total Fossil Fuels | Coal | Petroleum | Natural Gas (dry) | Total Solar Energy | Fuel Wood | Work Animal Feed | Direct Wind and Water Power (feed equivalent) | | Total Nuclear Energy |
	1	2	3	4	5	6	7	8	9	10	11
1970	100	94.29	18.87	43.25	32.17	5.37	1.31	0.16	0.03	3.87	0.33
1969	100	94.15	19.28	43.04	31.83	5.63	1.38	0.19	0.03	4.03	0.22
1968	100	94.33	20.14	43.04	31.15	5.46	1.46	0.23	0.04	3.73	0.21
1967	100	94.04	20.65	42.66	30.73	5.83	1.57	0.27	0.04	3.95	0.13
1966	100	94.31	21.71	42.38	30.22	5.59	1.63	0.31	0.05	3.60	0.10
1965	100	93.99	21.84	42.63	29.52	5.94	1.74	0.36	0.06	3.78	0.07
1964	100	94.00	21.48	42.68	29.84	5.93	1.83	0.40	0.06	3.64	0.07
1963	100	94.00	21.20	43.43	29.37	5.93	1.92	0.45	0.07	3.49	0.07
1962	100	93.63	20.93	43.69	29.01	6.32	2.01	0.50	0.07	3.74	0.05
1961	100	93.60	21.26	43.96	28.38	6.36	2.12	0.55	0.08	3.61	0.04
1960	100	93.50	22.10	43.73	27.67	6.49	2.18	0.61	0.09	3.61	0.01
1959	100	93.40	22.05	44.39	26.96	6.60	2.28	0.65	0.09	3.58	—
1958	100	93.00	22.86	44.61	25.53	7.00	2.38	0.72	0.10	3.80	—

1957	100	93.13	25.90	43.07	24.16	6.87	2.41	0.76	0.11	3.59
1956	100	93.18	27.23	43.16	22.79	6.82	2.44	0.82	0.11	3.45
1955	100	92.93	28.00	42.53	22.40	7.07	2.59	0.94	0.12	3.42
1954	100	92.18	26.95	42.64	22.59	7.82	2.87	1.14	0.14	3.67
1953	100	92.13	30.31	41.03	20.79	7.87	2.82	1.25	0.14	3.66
1952	100	91.53	31.07	40.14	20.32	8.47	2.94	1.47	0.14	3.92
1951	100	91.45	34.24	38.44	18.77	8.55	2.96	1.68	0.15	3.76
1950	100	90.69	35.94	37.54	17.11	9.41	3.24	2.00	0.16	4.01
1945	100	88.83	47.21	29.88	11.74	11.17	3.73	3.25	0.24	3.95
1940	100	86.48	47.05	29.20	10.23	13.52	5.10	4.99	0.33	3.10
1935	100	82.79	48.14	25.71	8.94	17.21	6.32	6.93	0.57	3.39
1930	100	84.07	53.32	23.05	7.70	15.93	5.69	6.84	0.64	2.76
1925	100	81.53	59.36	17.28	4.89	18.47	6.19	8.64	1.09	2.55
1920	100	78.54	64.06	11.06	3.42	21.46	6.65	10.36	1.56	2.89
1915	100	74.43	64.31	6.86	3.26	25.57	8.17	12.75	1.64	3.01
1910	100	73.89	65.87	5.22	2.80	26.11	9.14	12.90	1.56	2.51
1905	100	69.59	63.37	3.86	2.36	30.41	11.68	14.68	1.85	2.20
1900	100	60.58	56.60	1.89	2.09	39.42	16.68	18.54	2.34	1.86
1895	100	51.84	48.74	1.65	1.45	48.16	22.70	21.81	2.85	0.80
1890	100	48.08	43.64	1.68	2.76	51.92	27.02	21.51	3.17	0.22
1885	100	38.83	37.23	0.52	1.08	61.17	35.17	21.99	4.01	—
1880	100	31.71	30.29	1.42	—	68.29	42.05	21.58	4.66	—
1875	100	24.50	24.31	0.19	—	75.50	48.48	21.79	5.23	—
1870	100	19.74	19.53	0.21	—	80.26	53.92	20.67	5.67	—
1865	100	13.35	13.14	0.21	—	86.65	57.51	22.57	6.57	—
1860	100	11.44	11.37	0.07	—	88.56	57.99	23.37	7.20	—
1855	100	10.55	10.55	—	—	89.45	59.84	22.52	7.09	—
1850	100	6.58	6.58	—	—	93.42	64.22	22.05	7.15	—

during combustion. The values for nuclear fuel are the amounts of heat generated by nuclear fission in electric power plants. Direct windpower and direct waterpower are converted to an equivalent amount of animal feed that would have been required to generate the same work output by means of work animals, and hydroelectric power is converted to an equivalent amount of fuel that would have been required to generate the same electric power output in fossil steam plants.

Table A2.2 gives the annual fuel inputs contributed by the various sources as percentages of the total fuel input, so that the relative contribution of each source can be appreciated more easily.

TABLE A2.3 CONSUMPTION OF FOSSIL FUELS, SOLAR ENERGY, AND NUCLEAR ENERGY FOR ENERGY USES; AND OF FOSSIL FUELS FOR NONENERGY USES: THE UNITED STATES, 1850–1970

| | 1 C = 10^{16} Btu | | | Millions of Btu per Capita | |
| | Energy Uses | Nonenergy Uses | Millions Population | Energy Uses | Nonenergy Uses |
Year	1	2	3	4	5
1970	6.4537	0.3935	203.81	316.7	19.3
1969	6.2212	0.3826	201.38	308.9	19.0
1968	5.9483	0.3369	199.40	298.3	16.9
1967	5.6259	0.3124	197.46	284.9	15.8
1966	5.4724	0.2836	195.57	279.8	14.5
1965	5.1886	0.2635	193.53	268.1	13.6
1964	4.9915	0.2530	191.14	261.1	13.2
1963	4.8043	0.2496	188.48	254.9	13.2
1962	4.6312	0.2367	185.77	249.3	12.7
1961	4.4343	0.2260	182.99	242.3	12.4
1960	4.3648	0.2239	179.98	242.5	12.4
1959	4.2348	0.2135	177.14	239.1	12.1
1958	4.1128	0.1947	174.15	236.2	11.2
1957	4.1172	0.1946	171.19	240.5	11.4
1956	4.1251	0.1905	168.09	245.4	11.3
1955	3.9506	0.1702	165.07	239.3	10.3
1954	3.6227	0.1605	161.88	223.8	9.9
1953	3.7747	0.1490	158.96	237.5	9.4
1952	3.6656	0.1543	156.39	234.4	9.9
1951	3.7024	0.1599	153.98	240.4	10.4
1950	3.4515	0.1417	151.87	227.3	9.3

TABLE A2.3   (Continued)

Year	$1 \text{ C} = 10^{16}$ Btu		Millions of Btu per Capita		
	Energy Uses 1	Nonenergy Uses 2	Millions Population 3	Energy Uses 4	Nonenergy Uses 5
1945	3.2700	0.1132	133.43	245.1	8.5
1940	2.5795	0.0848	132.46	194.7	6.4
1935	2.1460	0.0629	127.25	168.6	4.9
1930	2.4910	0.0670	123.08	202.4	5.4
1925	2.4259	0.0515	115.83	209.4	4.4
1920	2.3823	0.0378	106.47	223.8	3.6
1915	2.0412	0.0258	100.55	203.0	2.6
1910	1.9086	0.0215	92.41	206.5	2.3
1905	1.5633	0.0150	83.82	186.5	1.8
1900	1.1998	0.0088	76.09	157.7	1.2
1895	1.0099	0.0057	69.58	145.1	0.8
1890	0.9249	0.0058	63.06	146.7	0.9
1885	0.7601	0.0027	56.66	134.2	0.5
1880	0.6760	0.0020	50.26	134.5	0.4
1875	0.5915	0.0009	45.07	131.2	0.2
1870	0.5359	0.0006	39.90	134.3	0.2
1865	0.4807	0.0004	35.70	134.6	0.1
1860	0.4551	0.0003	31.51	144.4	0.1
1855	0.3990	0.0002	27.39	145.7	0.1
1850	0.3228	0.0001	23.26	143.1	0.0

Table A2.3 separates out and tabulates the input of fossil fuels used for nonenergy purposes and also tabulates the aggregate input from all sources used strictly for energy.

Tables A2.4 through A2.7 collect and organize the basic data. Footnotes and sources for all tables are listed at the end of the appendix.

The data in this appendix have been drawn from the following sources:

1. U.S. Bureau of the Census, *Historical Statistics of the United States, Colonial Times to 1957,* U.S. Government Printing Office, Washington, D.C., 1960; in conjunction with *Historical Statistics of the United States, Colonial Times to 1957; Con-*

TABLE A2.4 ANNUAL CONSUMPTION OF FOSSIL FUELS: THE UNITED
STATES, 1850–1956

[In units of 1 C = 10^{16} Btu]

		Fossil Fuel Consumption			
Year	Total 1	Anthracite Coal 2	Bituminous Coal 3	Petroleum 4	Natural Gas (dry) 5
1956	4.0409	0.0610	1.1338	1.8627	0.9834
1955	3.8459	0.0599	1.1104	1.7524	0.9232
1954	3.4881	0.0683	0.9512	1.6132	0.8554
1953	3.6147	0.0711	1.1182	1.6098	0.8156
1952	3.4962	0.0897	1.0971	1.5334	0.7760
1951	3.5321	0.0940	1̇.2285	1.4848	0.7248
1950	3.2552	0.1013	1.1900	1.3489	0.6150
1949	3.0039	0.0958	1.1673	1.2119	0.5289
1948	3.2487	0.1275	1.3622	1.2557	0.5033
1947	3.1411	0.1224	1.4302	1.1367	0.4518
1946	2.9048	0.1369	1.3110	1.0480	0.4089
1945	3.0055	0.1311	1.4661	1.0110	0.3973
1944	3.0434	0.1509	1.5447	0.9703	0.3775
1943	2.9095	0.1450	1.5557	0.8607	0.3481
1942	2.6720	0.1435	1.4149	0.8034	0.3102
1941	2.5650	0.1338	1.2893	0.8568	0.2851
1940	2.3042	0.1245	1.1290	0.7781	0.2726
1939	2.0753	0.1262	0.9854	0.7098	0.2539
1938	1.9008	0.1148	0.8811	0.6701	0.2348
1937	2.1869	0.1280	1.1286	0.6835	0.2468
1936	2.0594	0.1351	1.0697	0.6325	0.2221
1935	1.8288	0.1298	0.9336	0.5680	0.1974
1934	1.7225	0.1410	0.9008	0.4988	0.1819
1933	1.6177	0.1260	0.8323	0.4994	0.1600
1932	1.5671	0.1283	0.8041	0.4753	0.1594
1931	1.8112	0.1484	0.9743	0.5170	0.1715
1930	2.1506	0.1718	1.1921	0.5898	0.1969
1929	2.2911	0.1815	1.3612	0.5542	0.1942
1928	2.1492	0.1871	1.3069	0.4964	0.1588
1927	2.1013	0.1897	1.3095	0.4556	0.1465
1926	2.1730	0.1961	1.3954	0.4480	0.1335
1925	2.0198	0.1627	1.3079	0.4280	0.1212
1924	1.9768	0.2050	1.2681	0.3867	0.1170

TABLE A2.4   (Continued)

Year	Fossil Fuel Consumption				
	Total 1	Anthracite Coal 2	Bituminous Coal 3	Petroleum 4	Natural Gas (dry) 5
1923	2.0958	0.2208	1.3598	0.4120	0.1032
1922	1.6540	0.1443	1.1185	0.3127	0.0785
1921	1.5754	0.2082	1.0266	0.2724	0.0682
1920	1.9007	0.2179	1.3325	0.2676	0.0827
1919	1.6792	0.2113	1.1688	0.2198	0.0793
1918	1.9686	0.2385	1.4588	0.1942	0.0771
1917	1.8842	0.2378	1.3835	0.1779	0.0850
1916	1.7052	0.2106	1.2631	0.1508	0.0807
1915	1.5385	0.2160	1.1134	0.1418	0.0673
1914	1.4858	0.2198	1.0703	0.1325	0.0632
1913	1.6074	0.2207	1.2034	0.1213	0.0620
1912	1.5093	0.2038	1.1402	0.1059	0.0594
1911	1.4027	0.2197	1.0245	0.1041	0.0544
1910	1.4261	0.2060	1.0654	0.1007	0.0540
1909	1.3018	0.1978	0.9685	0.0844	0.0511
1908	1.1762	0.2037	0.8478	0.0820	0.0427
1907	1.3390	0.2098	1.0079	0.0781	0.0432
1906	1.1507	0.1748	0.8793	0.0555	0.0411
1905	1.0983	0.1910	0.8091	0.0610	0.0372
1904	0.9816	0.1797	0.7155	0.0534	0.0330
1903	0.9924	0.1843	0.7315	0.0449	0.0317
1902	0.8426	0.1030	0.6733	0.0364	0.0299
1901	0.7996	0.1657	0.5808	0.0250	0.0281
1900	0.7322	0.1410	0.5431	0.0229	0.0252
1895	0.5265	0.1439	0.3511	0.0168	0.0147
1890	0.4475	0.1159	0.2903	0.0156	0.0257
1885	0.2962	0.0957	0.1883	0.0040	0.0082
1880	0.2150	0.0717	0.1337	0.0096	—
1875	0.1451	0.0578	0.0862	0.0011	—
1870	0.1059	0.0503	0.0545	0.0011	—
1865	0.0642	0.0304	0.0328	0.0010	—
1860	0.0521	0.0275	0.0243	0.0003	—
1855	0.0421	0.0216	0.0205	—	—
1850	0.0219	0.0109	0.0110	—	—

TABLE A2.5 ANNUAL CONSUMPTION OF FOSSIL FUELS: THE UNITED STATES, 1947–1970

[in units of 1C = 10^{16} Btu]

	Total		Anthracite Coal		Fossil Fuel Consumption Bituminous Coal and Lignite		Petroleum		Natural Gas (dry)	
Year	Energy	Non-energy	Energy	Non-energy	Energy	Non-energy	Energy	Non-energy	Energy	Non-energy
	1	2	3	4	5	6	7	8	9	10
1970	6.0630	0.3935	0.0210	—	1.2561	0.0151	2.6517	0.3097	2.1342	0.0687
1969	5.8349	0.3826	0.0224	—	1.2357	0.0152	2.5462	0.2960	2.0306	0.0714
1968	5.5922	0.3369	0.0258	—	1.2250	0.0151	2.4289	0.2763	1.9125	0.0455
1967	5.2717	0.3124	0.0274	—	1.1848	0.0134	2.2806	0.2529	1.7789	0.0461
1966	5.1446	0.2836	0.0290	—	1.2070	0.0135	2.1989	0.2406	1.7097	0.0295

1965	4.8612	0.2635	0.0328	—	1.1441	0.0139	2.1038	0.2204	1.5805	0.0292
1964	4.6769	0.2530	0.0365	—	1.0769	0.0130	2.0284	0.2103	1.5351	0.0297
1963	4.5011	0.2496	0.0361	—	1.0234	0.0119	1.9871	0.2079	1.4545	0.0298
1962	4.3210	0.2367	0.0363	—	0.9704	0.0122	1.9333	0.1934	1.3810	0.0311
1961	4.1361	0.2260	0.0404	—	0.9381	0.0121	1.8686	0.1801	1.2890	0.0338
1960	4.0667	0.2239	0.0447	—	0.9568	0.0125	1.8325	0.1742	1.2327	0.0372
1959	3.9413	0.2135	0.0478	—	0.9221	0.0111	1.8111	0.1636	1.1603	0.0388
1958	3.8112	0.1947	0.0483	—	0.9249	0.0117	1.7762	0.1452	1.0618	0.0378
1957	3.8208	0.1946	0.0528	—	1.0491	0.0149	1.7170	0.1400	1.0019	0.0397
1956	3.8307	0.1905	0.0610	—	1.0996	0.0146	1.7252	0.1373	0.9449	0.0386
1955	3.6594	0.1702	0.0599	—	1.0792	0.0149	1.6333	0.1191	0.8870	0.0362
1954	3.3270	0.1605	0.0683	—	0.9389	0.0123	1.5000	0.1132	0.8198	0.0350
1953	3.4659	0.1490	0.0711	—	1.1035	0.0147	1.5068	0.1031	0.7845	0.0312
1952	3.3419	0.1543	0.0897	—	1.0846	0.0125	1.4297	0.1037	0.7379	0.0381
1951	3.3722	0.1599	0.0940	—	1.2142	0.0143	1.3833	0.1015	0.6807	0.0441
1950	3.1134	0.1417	0.1013	—	1.1771	0.0129	1.2625	0.0863	0.5725	0.0425
1949	2.8707	0.1333	0.0958	—	1.1564	0.0109	1.1339	0.0781	0.4846	0.0443
1948	3.1057	0.1432	0.1275	—	1.3501	0.0121	1.1746	0.0813	0.4535	0.0498
1947	3.0354	0.1356	0.1224	—	1.4477	0.0123	1.0637	0.0731	0.4016	0.0502

tinuation to 1962 and Revisions, U.S. Government Printing Office, Washington, D.C., 1965.

2. Walter G. Dupree, Jr. and James A. West, *United States Energy Through the Year 2000,* U.S. Department of the Interior, Washington, D.C., December, 1972.

3. U.S. Bureau of the Census, *Statistical Abstract of the United States: 1969,* U.S. Government Printing Office, Washington, D.C., 1969.

4. U.S. Bureau of the Census, *Statistical Abstract of the United*

TABLE A2.6 ENERGY CONTENT OF WORK ANIMAL FEED: THE UNITED STATES, 1850–1970

[In units of $1C = 10^{16}$ Btu]

| Year | Work Animals, Including Young Animals (millions) | | | | |
	Farm Horses and Mules 1	Nonfarm Horses and Mules 2	Oxen 3	Total Animals 4	Feed Energy 5
1970	(1.20)	(−)	—	1.20	0.0108
1969	(1.40)	(−)	—	1.40	0.0126
1968	1.61	(−)	—	1.61	0.0145
1967	(1.80)	(−)	—	1.80	0.0162
1966	(2.00)	(−)	—	2.00	0.0180
1965	2.20	(−)	—	2.20	0.0198
1964	(2.36)	(−)	—	2.36	0.0212
1963	(2.52)	(−)	—	2.52	0.0227
1962	(2.68)	(−)	—	2.68	0.0241
1961	(2.84)	(−)	—	2.84	0.0256
1960	3.09	(−)	—	3.09	0.0278
1959	3.19	(−)	—	3.19	0.0287
1958	3.42	(−)	—	3.42	0.0308
1957	3.63	(−)	—	3.63	0.0327
1956	3.96	(−)	—	3.96	0.0356
1955	4.31	(−)	—	4.31	0.0388
1954	4.79	(−)	—	4.79	0.0431
1953	5.40	(0.05)	—	5.45	0.0490
1952	6.15	(0.10)	—	6.25	0.0562
1951	7.04	(0.15)	—	7.19	0.0647
1950	7.78	0.20	—	7.98	0.0718

TABLE A2.6   (Continued)

Year	Work Animals, Including Young Animals (millions)				
	Farm Horses and Mules 1	Nonfarm Horses and Mules 2	Oxen 3	Total Animals 4	Feed Energy 5
1945	11.95	(0.25)	—	12.20	0.1098
1940	14.48	0.30	—	14.78	0.1330
1935	16.68	(0.32)	—	17.00	0.1530
1930	19.12	0.33	—	19.45	0.1750
1925	22.57	(1.23)	—	23.80	0.2142
1920	25.74	2.13	—	27.87	0.2508
1915	26.49	(2.79)	—	29.28	0.2635
1910	24.21	3.45	—	27.66	0.2489
1905	22.08	(3.28)	(0.40)	25.76	0.2318
1900	21.00	3.11	0.79	24.90	0.2241
1895	20.56	(2.85)	(1.20)	24.61	0.2215
1890	18.05	2.59	1.60	22.24	0.2002
1885	14.80	(2.33)	(1.50)	18.63	0.1677
1880	12.78	2.07	1.41	16.26	0.1463
1875	10.88	(1.81)	(1.65)	14.34	0.1291
1870	8.88	1.55	1.89	12.32	0.1109
1865	(8.14)	(1.37)	(2.56)	12.07	0.1086
1860	7.40	1.19	3.23	11.82	0.1064
1855	(6.15)	(1.01)	(2.83)	9.99	0.0899
1850	4.90	0.83	2.43	8.16	0.0734

States: 1972, U.S. Government Printing Office, Washington, D.C., 1972.

5. J. Frederic Dewhurst and Associates, *America's Needs and Resources: A New Survey,* The Twentieth Century Fund, New York, 1955.

Notes and references for the tables are as follows:

Table A2.1.

   Columns 3, 4, and 5 are taken from Table A2.4 for the years 1850–1945 and from Table A2.5 for the years 1950–1970.

TABLE A2.7 CONSUMPTION OF ENERGY
FROM DIRECT WINDPOWER AND DIRECT
WATERPOWER: THE UNITED STATES, 1850–
1970

[In units of $1C = 10^{16}$ Btu]

Year	Direct Wind and Water Power (feed equivalent)	Year	Direct Wind and Water Power (feed equivalent)
1970	(0.0020)	1945	0.0081
1969	(0.0022)	1940	0.0088
1968	(0.0024)	1935	(0.0126)
1967	(0.0026)	1930	0.0163
1966	(0.0028)	1925	(0.0270)
1965	(0.0030)	1920	0.0378
1964	(0.0032)	1915	(0.0340)
1963	(0.0034)	1910	0.0301
1962	(0.0036)	1905	(0.0292)
1961	(0.0038)	1900	0.0283
1960	(0.0040)	1895	(0.0289)
1959	(0.0042)	1890	0.0295
1958	(0.0044)	1885	(0.0306)
1957	(0.0046)	1880	0.0316
1956	(0.0048)	1875	(0.0310)
1955	(0.0050)	1870	0.0304
1954	(0.0052)	1865	(0.0316)
1953	(0.0054)	1860	0.0328
1952	(0.0055)	1855	(0.0283)
1951	(0.0057)	1850	0.0238
1950	0.0059		

Column 7 is taken from Reference 1, Series M87 for the years
1850–1950 and for 1955. Values for 1951–1954 are interpolated, and values for 1956–1970 are estimated.
Column 8 is taken from Table A2.6.
Column 9 is taken from Table A2.7.
Column 10 is taken from Table A2.5 for the years 1950–1970, and
(with modification amounting to a 10 percent reduction in
value) from Table A2.4 for the years 1850–1945. In the years

for which Tables A2.4 and A2.5 overlap they differ
systematically, the values in Table A2.4 being somewhat smaller
than those in Table A2.5. In order that the two series should
join smoothly in Table A2.1, the values in Table A2.4 were
reduced by 10 percent before being entered in Table A2.1.
Column 11 is taken from Reference 2, Appendix A, Table 1.

Table A2.2 is derived from Table A2.1.

Table A2.3.
    Columns 1 and 2 for the years 1950–1970 are derived from data in
    Tables A2.4 and A2.5. By noting in Table A2.5 that nonenergy
    uses of bituminous coal and lignite run about 1 percent of total
    use, and that nonenergy uses of hydrocarbons run about 7
    percent of total use, and by assuming these proportions to hold
    in earlier years, it was possible to decompose the fossil-fuel input
    figures in Table A2.4 to reflect energy uses and nonenergy uses
    of fossil fuels, and to complete columns 1 and 2 for the years
    1850–1945.
    Column 3 is from Reference 4, Table 2 for the years 1900–1970, and
    from Reference 1, Series A2 for the years 1850–1895.

Table A2.4 is taken from Reference 1, Series M79-M84.

Table A2.5 is taken from Appendix A in Reference 2.

Table A2.6.
    Column 1 is taken from Reference 1, Series K201 and K203 for the
    years 1870–1960; from Reference 3, Table 751 for 1965 and
    1968; and from Reference 5, page 1103 for 1850 and 1860.
    Figures in parentheses are interpolated or estimated.
    Column 2 is taken from Reference 5, page 1103. Figures in
    parentheses are interpolated or estimated.
    Column 3 is based on data for the number of working oxen given in
    Reference 5, page 1103, assuming that work stock amounted to
    70 percent of all oxen.
    Column 5 is computed at 90 million Btu per year for each animal in
    Column 4. This figure is based on an average weight of 1500
    pounds per animal (including young animals), an average daily
    consumption of 1 kilogram of feed per 100 pounds of animal,
    and an average energy content of 4 kilocalories per gram of feed.

Table A2.7. The work output of windmills, sailing vessels and direct-drive waterwheels is taken from Reference 5, page 1114. Assuming 4 percent as
the average efficiency of conversion of feed energy input to work output
by nonfarm work animals, the work output of direct windpower and
direct waterpower was multiplied by 25 before being entered in Table
A2.7. Figures in parentheses are interpolated or estimated. If a work
animal works 8 hours per day, 7 days per week, 50 weeks per year, at
3/4 horsepower average output, this amounts to 2100 horsepower-hour
per year, or 5.3 million Btu per year per working animal. If 70 percent
of all animals are workstock, the work output is 3.7 million Btu per
year per total animal. With an input of 90 million Btu per year per total
animal, this amounts to about 4 percent efficiency.

APPENDIX 3

PROJECTIONS OF UNITED STATES

ENERGY CONSUMPTION

The discussion in Chapter 1 led to a model of energy consumption in which the basic energy used in commerce, industry, and transportation is proportional to the product of three factors:

$$\left(\frac{\text{nonfarm working population}}{\text{total population}}\right) \times \left(\begin{array}{c}\text{energy content} \\ \text{per unit of} \\ \text{goods and services}\end{array}\right) \times$$

$$\text{(total population)}.$$

The discussion in Chapter 3 led to specific assumptions regarding the future growth rates anticipated for the first two factors. In summary, these assumptions are as follows:

Factor	1960–1968 Annual Growth Rate (%)	Assumptions Regarding Future Growth Rate
$\left(\dfrac{\text{nonfarm working population}}{\text{total population}}\right)$	1.6	Declines to zero by 2000 Overall factor 1.22 relative to 1968
$\left(\begin{array}{c}\text{energy content} \\ \text{per unit of} \\ \text{goods and services}\end{array}\right)$	0.9	Declines to zero by 2010 Overall factor 1.19 relative to 1968

These factors are tabulated year by year in Table A3.1 following the stated assumptions concerning future growth. Their product, also tabulated, is to be viewed as the per capita energy projected to be consumed by commerce, industry, and transportation for their basic energy uses, relative to 1968 = 1. Although individual end uses classed as basic uses in commerce, industry, and transportation grew at somewhat different rates during 1960–1968, they are projected to grow at equal rates in the future, following the per capita trajectory in Table A3.1. As a specific example, consider industrial direct heat as a

TABLE A3.1 PROJECTION OF ENERGY FOR BASIC USES IN COMMERCE, INDUSTRY, AND TRANSPORTATION

Year	(1)	(2)	(3)
1968	1.000	1.0000	1.000
1969	1.014	1.0088	1.023
1970	1.027	1.0175	1.045
1971	1.039	1.0260	1.066
1972	1.051	1.0343	1.087
1973	1.063	1.0425	1.108
1974	1.074	1.0505	1.128
1975	1.085	1.0584	1.148
1976	1.096	1.0660	1.168
1977	1.106	1.0735	1.187
1978	1.116	1.0808	1.206
1979	1.125	1.0879	1.224
1980	1.134	1.0948	1.242
1981	1.142	1.1015	1.258
1982	1.150	1.1080	1.274
1983	1.158	1.1143	1.290
1984	1.165	1.1203	1.305
1985	1.172	1.1261	1.320
1986	1.178	1.1317	1.333
1987	1.184	1.1371	1.346
1988	1.189	1.1422	1.358
1989	1.194	1.1471	1.370
1990	1.198	1.1518	1.380
1991	1.202	1.1563	1.389
1992	1.206	1.1605	1.398

TABLE A3.1   (Continued)

Year	(1)	(2)	(3)
1993	1.209	1.1644	1.407
1994	1.212	1.1681	1.416
1995	1.215	1.1716	1.424
1996	1.217	1.1748	1.430
1997	1.218	1.1777	1.435
1998	1.219	1.1804	1.439
1999	1.220	1.1829	1.443
2000	1.220	1.1850	1.446
2001	1.220	1.1869	1.448
2002	1.220	1.1885	1.450
2003	1.220	1.1899	1.452
2004	1.220	1.1910	1.453
2005	1.220	1.1918	1.454
2006	1.220	1.1924	1.454
2007	1.220	1.1928	1.455
2008	1.220	1.1928	1.455

$$(1) = \frac{\text{nonfarm working population}}{\text{total population}}.$$

(2) = Energy content per unit of goods and services.

(3) = (1) $\times$ (2) = per capita energy projected to be consumed by commerce, industry, and transportation for their basic energy uses, relative to 1968 = 1.

basic industrial energy use. In 1968 this use consumed 7,007 trillion Btu of energy as indicated in Table A1.2 of Appendix 1. If we want to know how much industrial direct heat is projected to be consumed in 1985, we look in the last column of Table A3.1 opposite 1985 and find the factor 1.32. Multiplying the 1968 figure of 7,007 trillion Btu by 1.32 we obtain 9,249 trillion Btu as the projected consumption on the basis of a static population. If we assume population growth at 1.3 percent annually from 1968 to 1985, we must multiply again by the population growth factor $(1.013)^{17} = 1.246$, obtaining 11,-524 Btu.

The model of energy consumption in Chapter 1 considered that residential basic energy consumption was proportional to the product

$$\left(\frac{\text{residential basic energy}}{\text{total population}}\right) \times (\text{total population})$$

and that new energy consumption was proportional to the product

$$\left(\frac{\text{new energy}}{\text{total population}}\right) \times (\text{total population}).$$

TABLE A3.2 PROJECTIONS OF ENERGY FACTORS FOR KEY CONSUMPTION SEGMENTS, RELATIVE TO 1968 = 1

Year	(1)	(2)	(3)
1968	1.000	1.000	1.00
1969	1.023	1.027	1.08
1970	1.045	1.052	1.15
1971	1.066	1.076	1.23
1972	1.087	1.101	1.30
1973	1.108	1.124	1.37
1974	1.128	1.146	1.44
1975	1.148	1.167	1.51
1976	1.168	1.186	1.57
1977	1.187	1.204	1.63
1978	1.206	1.221	1.69
1979	1.224	1.237	1.74
1980	1.242	1.250	1.79
1981	1.258	1.263	1.83
1982	1.274	1.273	1.87
1983	1.290	1.282	1.90
1984	1.305	1.289	1.93
1985	1.320	1.295	1.95
1986	1.333	1.298	1.97
1987	1.346	1.300	1.98
1988	1.358	1.300	1.99
1989	1.370	1.300	2.00
1990	1.380	1.300	2.00

TABLE A3.2   (Continued)

Year	(1)	(2)	(3)
1991	1.389	1.300	2.00
1992	1.398	1.300	2.00
1993	1.407	1.300	2.00
1994	1.416	1.300	2.00
1995	1.424	1.300	2.00
1996	1.430	1.300	2.00
1997	1.435	1.300	2.00
1998	1.439	1.300	2.00
1999	1.443	1.300	2.00
2000	1.446	1.300	2.00
2001	1.448	1.300	2.00
2002	1.450	1.300	2.00
2003	1.452	1.300	2.00
2004	1.453	1.300	2.00
2005	1.454	1.300	2.00
2006	1.454	1.300	2.00
2007	1.455	1.300	2.00
2008	1.455	1.300	2.00

(1) Per capita basic energy in commerce, industry, and
transportation from Column 3 of Table A3.1.
(2) Per capita residential basic energy.
(3) Per capita new energy.

The discussion in Chapter 3 led to specific assumptions regarding the future
growth rates anticipated for the left-hand factors as follows:

Factor	1960–1968 Annual Growth Rate (%)	Assumptions Regarding Future Growth Rate
$\left(\dfrac{\text{residential basic energy}}{\text{total population}}\right)$	2.8	Declines to zero by 1990 Overall factor 1.30 relative to 1968
$\left(\dfrac{\text{new energy}}{\text{total population}}\right)$	7.2	Declines to zero by 1990 Overall factor 2.0 relative to 1968

TABLE A3.3 PROJECTION OF PER CAPITA
ENERGY CONSUMPTION IN THE UNITED
STATES, RELATIVE TO 1968 = 1

Year	Consumption	Year	Consumption
1968	1.000	1988	1.416
1969	1.029	1989	1.426
1970	1.057	1990	1.433
1971	1.085	1991	1.440
1972	1.112	1992	1.446
1973	1.138	1993	1.453
1974	1.164	1994	1.460
1975	1.189	1995	1.465
1976	1.213	1996	1.470
1977	1.236	1997	1.474
1978	1.260	1998	1.477
1979	1.280	1999	1.479
1980	1.301	2000	1.482
1981	1.319	2001	1.483
1982	1.337	2002	1.485
1983	1.353	2003	1.486
1984	1.369	2004	1.487
1985	1.383	2005	1.488
1986	1.395	2006	1.488
1987	1.406	2007	1.488

Table A3.2 repeats the basic energy factor for commerce, industry and transportation from Table A3.1 and shows in adjacent columns the basic residential energy factor and the new energy factor following the assumptions just stated for them.

Now it is possible to make projections of energy consumption and electricity consumption corresponding to various assumptions about the future. First let us project total energy consumption, per capita, relative to 1968. In 1968 the proportions of energy consumption in various key consumption segments were:

Basic energy in commerce, industry, and transportation	0.737
Residential basic energy	0.157
New energy	0.106
	1.000

The projection assumes that each of the above three energy segments grows at its characteristic pace as indicated by the factors in the corresponding columns of Table A3.2. The resulting overall growth is the sum of the three:

$$0.737 \times \text{(basic energy factor for commerce, industry and transportation)}$$
$$+$$
$$0.157 \times \text{(residential basic energy factor)}$$
$$+$$
$$0.106 \times \text{(new energy factor)}$$

This sum is tabulated in Table A3.3. It shows that United States per capita energy consumption is projected to level off at 1.488 times the 1968 level. This result has more significant figures than is warranted by the quality of the assumptions, because the calculation is carried out as though the assumptions were rigorously true. In the text the factor 1.488 is rounded to a more realistic 1 1/2.

It is possible also to project the consumption of electricity. As a lower bound for electricity consumption, we may assume that there is no change in the degree of electrification of any end use, and that there is no change in the efficiency of electric generation. Then the fuel input to electric generation for each end use grows along the appropriate trajectory tabulated in Table A3.2. With population growth at 1.3 percent annually, the resulting growth of fuel input and electric output from electric generation is shown in Figure 11.5 of Chapter 11. As an upper bound for electricity consumption we may assume full electrification in 2000 of all end uses except process steam and transportation. Based on the factors for 2000 in Table A3.2 this corresponds to the upper bound in Figure 11.5.

The factors tabulated in Table A3.2, and hence all projections derived from them, contain an implicit assumption that the progress of electrification will follow more or less an extension of its old trajectory. If electrification deviates too far above or below an extension of its past trend, as, for example, at the lower and upper bounds of electrification just considered, the factors in Table A3.2 should probably be changed somewhat, and the lower and upper bounds—which depend on the factors—would be changed somewhat too. I ignore these effects as being relatively small and in any event beyond the degree of accuracy hoped for in making these projections.

APPENDIX 4

FACTORS FOR CONVERTING

TO 1970 DOLLARS

Rather than using a price index such as the consumer price index or the wholesale price index, each of which combines prices of individual items with fixed weights and represents only a segment of the entire economy, I have used the implicit price deflator for GNP, a by-product of the calculation of real GNP, to convert all dollar figures to constant 1970 dollars in a manner that more fully reflects the overall economy. The construction of the deflator is explained in the article "Alternative Measures of Price Change for GNP" by Allan H. Young and Claudia Harkins, *Survey of Current Business,* March 1969. The implicit price deflator for gross national product (with 1958 = 100) has been published for the years 1929–1966 on page 52 of *Survey of Current Business,* September 1967, and the series has been continued year after year in subsequent issues of the *Survey.* For years prior to 1929, implicit price deflators (with 1929 = 100) are given in Series F5 of *Historical Statistics of the United States, Colonial Times to 1957,* U.S. Government Printing Office, Washington, D.C., 1960.

I have adjusted all of these price deflators to 1970 = 100, and have taken their reciprocals. The resulting reciprocal GNP deflator factors are given in the following tabulation. The reciprocal GNP deflator factor for any year is the number of 1970 dollars per dollar of that year. Its utility is as follows: to convert dollars for year x into 1970 dollars, multiply by the tabulated factor for year x.

183

Year	Factor for Converting to 1970 Dollars	Year	Factor for Converting to 1970 Dollars	Year	Factor for Converting to 1970 Dollars
1972	0.927	1947	1.814	1922	2.73
1971	0.955	1946	2.028	1921	2.60
1970	1.000	1945	2.266	1920	2.21
1969	1.055	1944	2.325	1919	2.52
1968	1.106	1943	2.382		
1967	1.151	1942	2.553		Department
1966	1.188	1941	2.866		of Commerce
1965	1.220	1940	3.082		concept
1964	1.243	1939	3.132	1917–1921	2.55
1963	1.262	1938	3.082	1912–1916	4.2
1962	1.279	1937	3.040	1907–1911	4.7
1961	1.293	1936	3.168	1902–1906	5.1
1960	1.310	1935	3.176	1897–1901	5.7
1959	1.332	1934	3.206	1892–1896	5.8
1958	1.353	1933	3.442	1889–1893	5.5
1957	1.388	1932	3.365		
1956	1.439	1931	3.020		Kuznets
1955	1.488	1930	2.744		concept
1954	1.510	1929	2.674	1889–1893	5.3
1953	1.532	1928	2.67	1887–1891	5.2
1952	1.546	1927	2.70	1882–1886	4.9
1951	1.580	1926	2.65	1877–1881	4.7
1950	1.687	1925	2.65	1872–1876	4.0
1949	1.710	1924	2.70	1869–1873	3.6
1948	1.700	1923	2.67		

APPENDIX 5

FIELD CONSTRUCTION COSTS VERSUS

FACTORY CONSTRUCTION COSTS
(AN ILLUSTRATIVE COST MODEL)

For many decades, the electric utility industry has followed a pattern of plant construction in which a portion of the construction was done in the field at the plant site (primarily the construction of the building and the boiler), and a portion was done in large factories supplying many utility plants (primarily the construction of the turbine, generator, and power conditioning equipment). As industry capacity and plant capacity doubled every decade, factory capacity also doubled, as did field construction at each plant site. Manufacturing and construction costs per kilowatt declined in the factory and in the field, since each of these activities increased its scale of operations. As long as both activities grew in proportion, the economies of scale produced similar cost reductions in each, and therefore an overall cost reduction, although the unit cost of field construction was always higher than the unit cost of factory construction. The pattern held until plant size reached about 200 megawatts.

Then analysis suggested to design engineers that the economy of scale was more important to nuclear plants than to fossil-fuel plants, so that nuclear power would be relatively cheaper in very large plants. Nuclear plant size was increased accordingly. However, with the construction of 1000-megawatt plants, a greater proportion of construction was shifted from the factory to the field, upsetting the pattern of the past, with the result that a greater proportion of the overall work of power plant manufacture or construction was carried out at smaller, less-efficient field locations. Since unit costs in the field run several times those in the factory, a shift of activity from factory to field can cause an increase in overall unit cost. An anticipated economy of scale may prove to be illusory.

This qualitative line of discussion can be illustrated by an idealized quantitative example. Assume that the cost of construction of 1 kilowatt of generation capacity declines by 20 percent every time factory production is doubled. Let some of the work be done in a large factory producing 10,000 megawatts of capacity per year. Let the balance of the work be done in the field where a 1000-megawatt plant is being built over a 5-year time span. Let the manufacturing cost per kilowatt in the large factory be taken as unity. The cost of field work will be higher, because the construction site can be viewed as a little factory where output is relatively small.

If all the work were done in the field, the production of that little factory would amount to 200 megawatts of capacity per year. If only a fraction (call it f, for fraction or field) of the work were done in the field, the production of the little plant-site factory would be $200 f$ megawatts of capacity per year. Factory production is larger than field production by the ratio

$$10,000/200f = 50/f.$$

Now consider the cost of a kilowatt of capacity as it depends on f. When $f = 0$, none of the work is done in the field. All the work is done in the factory, and the unit cost is 1, by definition of the standard cost. When $f = 1$, all the work is done in the field and the unit cost is 3.5, the appropriate value for a production rate 50 times smaller (the cost is multiplied by a factor $1/0.8 = 1.25$ every time production rate is halved, the converse of cost decreasing by a factor of 0.8 every time production rate is doubled). The cost components vary with f as follows:

f = Fraction of Manufacturing Work Done in the Field	Unit Cost of Factory Portion	Unit Cost of Field Portion	Total Unit Cost
0.0	1	—	1.00
0.1	1	7.31	1.63
0.2	1	5.85	1.97
0.3	1	5.14	2.24
0.4	1	4.69	2.48
0.5	1	4.36	2.68
0.6	1	4.12	2.87
0.7	1	3.92	3.04
0.8	1	3.75	3.20
0.9	1	3.62	3.36
1.0	1	3.50	3.50

The first bit of field construction hurts the most. Unit cost can be doubled by moving 20 percent of the work to the field from an otherwise pure factory operation. The moral of the analysis is crystal clear. Because of its relatively high cost, field construction must be kept at an absolute minimum. Insofar as possible, major modules must be fabricated, assembled, tested, and shipped intact, with field activity limited to connecting them up.

Although the analysis in this appendix is idealized and oversimplified, I believe that it illustrates one contribution to the higher-than-expected costs of nuclear plant construction. Nuclear steam plants are more expensive than fossil-fuel steam plants of the same rating because of additional shielding, containment, and other specifically nuclear features that they require, and because of additional heat-recovery equipment added to help compensate for the lower operating temperature. The larger the nuclear plant, the smaller the ratio of heat recovery costs and specifically nuclear costs to total cost. This is the motivation for increasing the size of nuclear plants. But because the proportion of field construction increased for these large plants, their cost turned out to be higher than anticipated.

INDEX